Bio-Based Solvents

Wiley Series in Renewable Resources

Series Editor

Christian V. Stevens – Faculty of Bioscience Engineering, Ghent University, Ghent, Belgium

Titles in the Series

Wood Modification: Chemical, Thermal and Other Processes
Callum A. S. Hill

Renewables-Based Technology: Sustainability Assessment
Jo Dewulf, Herman Van Langenhove

Biofuels
Wim Soetaert, Erik Vandamme

Handbook of Natural Colorants
Thomas Bechtold, Rita Mussak

Surfactants from Renewable Resources
Mikael Kjellin, Ingegärd Johansson

Industrial Applications of Natural Fibres: Structure, Properties and Technical Applications
Jörg Müssig

Thermochemical Processing of Biomass: Conversion into Fuels, Chemicals and Power
Robert C. Brown

Biorefinery Co-Products: Phytochemicals, Primary Metabolites and Value-Added Biomass Processing
Chantal Bergeron, Danielle Julie Carrier, Shri Ramaswamy

Aqueous Pretreatment of Plant Biomass for Biological and Chemical Conversion to Fuels and Chemicals
Charles E. Wyman

Bio-Based Plastics: Materials and Applications
Stephan Kabasci

Introduction to Wood and Natural Fiber Composites
Douglas D. Stokke, Qinglin Wu, Guangping Han

Cellulosic Energy Cropping Systems
Douglas L. Karlen

Introduction to Chemicals from Biomass, 2nd Edition
James H. Clark, Fabien Deswarte

Lignin and Lignans as Renewable Raw Materials: Chemistry, Technology and Applications
Francisco G. Calvo-Flores, Jose A. Dobado, Joaquín Isac-García, Francisco J. Martín-Martínez

Sustainability Assessment of Renewables-Based Products: Methods and Case Studies
Jo Dewulf, Steven De Meester, Rodrigo A. F. Alvarenga

Cellulose Nanocrystals: Properties, Production and Applications
Wadood Hamad

Fuels, Chemicals and Materials from the Oceans and Aquatic Sources
Francesca M. Kerton, Ning Yan

Forthcoming Titles

Biorefinery of Inorganics: Recovering Mineral Nutrients from Biomass and Organic Waste
Erik Meers, Gerard Velthof

Nanoporous Catalysts for Biomass Conversion
Feng-Shou Xiao, Liang Wang

The Chemical Biology of Plant Biostimulants
Danny Geelen

Bio-Based Packaging
Mohd Sapuan Salit, Muhammed Lamin Sanyang

Thermochemical Processing of Biomass: Conversion into Fuels, Chemicals and Power, 2nd Edition
Robert Brown

Bio-Based Solvents

Edited by

FRANÇOIS JÉRÔME
Institut de Chimie des Milieux et Matériaux de Poitiers
Université de Poitiers, ENSIP
France

RAFAEL LUQUE
Departamento de Química Orgánica
Universidad de Córdoba
Spain

WILEY

This edition first published 2017
© 2017 John Wiley & Sons Ltd

Registered Office(s)
John Wiley & Sons, Inc., 111 River Street, Hoboken, NJ 07030, USA
John Wiley & Sons Ltd, The Atrium, Southern Gate, Chichester, West Sussex, PO19 8SQ, UK

Editorial Office
9600 Garsington Road, Oxford, OX4 2DQ, UK

For details of our global editorial offices, customer services, and more information about Wiley products visit us at www.wiley.com.

Wiley also publishes its books in a variety of electronic formats and by print-on-demand. Some content that appears in standard print versions of this book may not be available in other formats.

Library of Congress Cataloging-in-Publication Data

Names: Jérôme, François, 1974- editor. | Luque, Rafael, editor.
Title: Bio-based solvents / edited by François Jérôme, National Higher
  Engineering School of Poitiers (ENSIP), University of Poitiers, France,
  Rafael Luque, Departament of Quimica Organica, University of Cordoba,
  Spain.
Other titles: Biobased solvents
Description: Hoboken, NJ : John Wiley & Sons, Inc., 2017. | Series: Wiley
  series in renewable resources | Includes bibliographical references and
  index.
Identifiers: LCCN 2017007717 (print) | LCCN 2017008295 (ebook) | ISBN
  9781119065395 (cloth) | ISBN 9781119065432 (pdf) | ISBN 9781119065449
  (epub)
Subjects: LCSH: Solvents. | Green chemistry.
Classification: LCC TP247.5 .B56 2017 (print) | LCC TP247.5 (ebook) | DDC
  660/.29482–dc23
LC record available at https://lccn.loc.gov/2017007717

Cover Design: Wiley
Cover Images: (Top Image) © herjua/Gettyimages; (Bottom Left) Ingram Publishing / Alamy Stock Photo

Set in 10/12pt, TimesLTStd by SPi Global, Chennai, India.
Printed and bound in Malaysia by Vivar Printing Sdn Bhd

10 9 8 7 6 5 4 3 2 1

Contents

List of Contributors

Paula Bracco   Biocatalysis, Department of Biotechnology, TU Delft, The Netherlands

Fergal Byrne   Green Chemistry Centre of Excellence, Department of Chemistry, University of York, UK

James H. Clark   Green Chemistry Centre of Excellence, Department of Chemistry, University of York, UK

Annelies Dewaele   Centre for Surface Chemistry and Catalysis, KU Leuven, Belgium

Pablo Domínguez de María   Sustainable Momentum SL, Las Palmas de Gran Canaria, Spain

Thomas J. Farmer   Green Chemistry Centre of Excellence, Department of Chemistry, University of York, UK

Amandine Foulet   Institut des Sciences Moléculaires, Université de Bordeaux, France

Joaquín García-Álvarez   CSIC, Laboratorio de Compuestos Organometálicos y Catálisis, Centro de Innovación en Química Avanzada, Universidad de Oviedo, Spain

Eskinder Gemechu   Institut des Sciences Moléculaires, Université de Bordeaux, France

Yanlong Gu   School of Chemistry and Chemical Engineering, Huazhong University of Science and Technology, China

Andrew J. Hunt   Green Chemistry Centre of Excellence, Department of Chemistry, University of York, UK

François Jérôme   CNRS, Institut de Chimie des Milieux et Matériaux de Poitiers, Université de Poitiers, ENSIP, France

Saimeng Jin   Green Chemistry Centre of Excellence, Department of Chemistry, University of York, UK

Yuhe Liao   Centre for Surface Chemistry and Catalysis, KU Leuven, Belgium

Philippe Loubet   Institut des Sciences Moléculaires, Université de Bordeaux, France

Rafael Luque   Departamento de Química Orgánica, Universidad de Córdoba, Campus de Rabanales, Spain

C. Rob McElroy   Green Chemistry Centre of Excellence, Department of Chemistry, University of York, UK

James Mgaya   Chemistry Department, University of Dar es Salaam, Tanzania

Egid B. Mubofu   Chemistry Department, University of Dar es Salaam, Tanzania

Joan J. E. Munissi   Chemistry Department, University of Dar es Salaam, Tanzania

Karine de Oliveira Vigier   CNRS, Institut de Chimie des Milieux et Matériaux de Poitiers, Université de Poitiers, France

Palanisamy Ravichandiran   School of Chemistry and Chemical Engineering, Huazhong University of Science and Technology, China

Bert F. Sels   Centre for Surface Chemistry and Catalysis, KU Leuven, Belgium

James Sherwood   Green Chemistry Centre of Excellence, Department of Chemistry, University of York, UK

Guido Sonnemann   Institut des Sciences Moléculaires, Université de Bordeaux, France

Michael Tsang   Institut des Sciences Moléculaires, Université de Bordeaux, France

Danny Verboekend   Centre for Surface Chemistry and Catalysis, KU Leuven, Belgium

Series Preface

Renewable resources, their use and modification are involved in a multitude of important processes with a major influence on our everyday lives. Applications can be found in the energy sector, chemistry, pharmacy, the textile industry, paints and coatings, to name but a few.

The area interconnects several scientific disciplines (agriculture, biochemistry, chemistry, technology, environmental sciences, forestry ...), which makes it very difficult to have an expert view on the complicated interaction. Therefore, the idea to create a series of scientific books, focusing on specific topics concerning renewable resources, has been very opportune and can help to clarify some of the underlying connections in this area.

In a very fast changing world, trends are not only characteristic for fashion and political standpoints; also, science is not free from hypes and buzzwords. The use of renewable resources is again more important nowadays; however, it is not part of a hype or a fashion. As the lively discussions among scientists continue about how many years we will still be able to use fossil fuels – opinions ranging from 50 to 500 years – they do agree that the reserve is limited and that it is essential not only to search for new energy carriers but also for new material sources.

In this respect, renewable resources are a crucial area in the search for alternatives for fossil-based raw materials and energy. In the field of energy supply, biomass and renewable-based resources will be part of the solution alongside other alternatives such as solar energy, wind energy, hydraulic power, hydrogen technology and nuclear energy.

In the field of material sciences, the impact of renewable resources will probably be even bigger. Integral utilization of crops and the use of waste streams in certain industries will grow in importance, leading to a more sustainable way of producing materials.

Although our society was much more (almost exclusively) based on renewable resources centuries ago, this disappeared in the Western world in the nineteenth century. Now it is time to focus again on this field of research. However, it should not mean a 'retour à la nature', but it should be a multidisciplinary effort on a highly technological level to perform research towards new opportunities, to develop new crops and products from renewable resources. This will be essential to guarantee a level of comfort for a growing number of people living on our planet. It is 'the' challenge for the coming generations of scientists to develop more sustainable ways to create prosperity and to fight poverty and hunger in the world. A global approach is certainly favoured.

This challenge can only be dealt with if scientists are attracted to this area and are recognized for their efforts in this interdisciplinary field. It is, therefore, also essential that consumers recognize the fate of renewable resources in a number of products.

Furthermore, scientists do need to communicate and discuss the relevance of their work. The use and modification of renewable resources may not follow the path of the genetic engineering concept in view of consumer acceptance in Europe. Related to this aspect, the series will certainly help to increase the visibility of the importance of renewable resources.

Being convinced of the value of the renewables approach for the industrial world, as well as for developing countries, I was myself delighted to collaborate on this series of books focusing on different aspects of renewable resources. I hope that readers become aware of the complexity, the interaction and interconnections, and the challenges of this field and that they will help to communicate on the importance of renewable resources.

I certainly want to thank the people of Wiley's Chichester office, especially David Hughes, Jenny Cossham and Lyn Roberts, in seeing the need for such a series of books on renewable resources, for initiating and supporting it and for helping to carry the project to the end.

Last, but not least, I want to thank my family, especially my wife Hilde and children Paulien and Pieter-Jan, for their patience and for giving me the time to work on the series when other activities seemed to be more inviting.

Christian V. Stevens, Faculty of Bioscience Engineering

Ghent University, Belgium

Series Editor 'Renewable Resources'

June 2005

Foreword

The present-day solvent market is of the order of 20 million tonnes and worth tens of billions of US dollars annually to the global economy. European solvent production provides about one-quarter of the worldwide market. The sheer volumes involved, the diversity of applications and the prevalence of small, functional compounds that often contain heteroatoms helps make the solvent sector a top candidate for switching to safer and more sustainable alternatives under the pressure of regional and global chemical regulation, notably REACh (Registration, Evaluation, Authorisation & restriction of Chemicals). A critical stage in the REACh process is imminent as the small- to medium-volume chemicals are registered in time for the 2018 deadline. As several commonly used solvents like NMP (*N*-methyl-2-pyrrolidone) are under close scrutiny at the time of writing, we can assume that the number of problematic solvents identified under REACh (and possibly other legislation) will be far greater at the time of reading.

The search for "greener" solvents is not new. If we go back to the early days of green chemistry in the 1990s, "alternative solvents" was one of the most popular research areas, with more and more articles reporting uses for known alternatives, including liquid and supercritical carbon dioxide, and an ever-increasing number of newly reported ionic liquids. These represented potentially positive step changes to chemical manufacturing technologies. Supercritical CO_2 enables rapid and easy separation after reaction (since separations are commonly a major contributor to the low environmental impact of many chemical processes) and ionic liquids can avoid the critical environmental concerns around using volatile organic compounds (these being threats to human health and causes of atmospheric damage). Research in these areas has continued, though few industrial processes have changed to incorporate these step-change technologies. The costs of such major changes to the processes, the added energy and capital expenditure costs of working with supercritical fluids, and the toxicity, separation and purification challenges associated with some ionic liquids have inhibited progress. Among the most likely ionic liquids to have a future in industrial chemistry are deep eutectic mixtures as well as other low-melting mixtures that are constructed from bio-based compounds. These are the subject of a book chapter here. Other "green solvent" approaches including greater use of water as a reaction solvent; and no-solvent processes have had some impact, but the vast majority of solvent applications have remained essentially unchanged. In the meantime, the rapidly growing number of synthetic transformations used by the pharmaceutical industry have effectively increased the breadth and complexity of the problems (e.g. more metal-catalysed processes and more processes that need polar aprotic solvents). Other, newer industries, such as advanced materials, are creating additional problems (e.g. the current use of solvents like NMP to process graphene). The need for safer, cost-effective solvents has never been greater.

Bio-based organic solvents are another way to make chemical processes more sustainable, and despite the infancy of the area of bio-based chemicals, the annual bio-based solvent use in the European Union is projected to grow to over one million tonnes by 2020.

In the European Union, for example, a strategy for implementing and encouraging a bio-based economy has been launched and a mandate issued specifically addressing the development of standards relating to bio-based solvents. As a tool to support and enhance the bio-based economy, the purpose of standards is to increase market transparency and establish common requirements for products in order to guarantee certain characteristics, such as a minimum value of bio-based content. Bio-based solvents must also compete economically with established petrochemical solvents in order to gain a significant market share. It is also important to note that standards for bio-based products will increasingly include considerations of feedstocks – their renewability and sustainability, as well as end-of-life issues, potentially extending to the recovery of resources consistent with the "circular economy". Life cycle assessments for greener solvents are described in a chapter in this book.

But what should future bio-based solvents look like? Is it sufficient for them to provide the advantages of sustainability and biodegradability? The problem with replacing petroleum-derived solvents with the same bio-based solvent is that any safety or toxicity issues are not resolved. Environmental issues occurring at the end of use will also persist. With the REACh European regulation starting to influence solvent selection, manufacturers will be forced to investigate alternative solvents. At least bio-based solvents are compatible with the development of environmentally sustainable processes. We must assume that new solvents will be needed to meet the highly demanding requirements of the current breadth of solvent properties. Nature provides few naturally occurring compounds that can act as solvents, though modern biotechnology enables access to large volumes of a number of useful small molecules, some of which can be directly used as solvents (e.g. ethanol) and others that can be easily converted into solvents (e.g. lactates from lactic acid). But the creation of new bio-based solvents with properties similar to many existing solvent types, including aromatics, halogenated solvents and amides, will be challenging. In this book we look at bio-based aromatic solvents in some detail.

Regarding the origin of bio-based solvents, it is important that bio-waste streams, including forestry wastes and food supply chain wastes (from farm to fork), should be the source of chemical products where at all possible. This is because two substantial issues detract from the advantages of solvent substitution in favour of "first-generation" sugar-derived bio-based solvents, especially those made by fermentation. This feedstock competes with our food supply, therefore creating a strongly objectionable conflict. Extending this argument, non-food crops for use in the chemical feedstock or biofuel sectors also require arable land, thus still creating pressure on food production (as well as biodiversity and other sustainability issues). Nonetheless, biofuels have quickly become a major part of the bio-economy, in regions from the Americas to Europe as well as in Asia and beyond. The success of the petrochemical industry is largely based on the availability of large quantities of inexpensive feedstock, and this has been enabled by the emergence and continued strength of the (petroleum) oil industry. We must learn to do the same in the bio-economy. The chapter on glycerol illustrates this by considering this major by-product from bio-diesel manufacturing as a solvent, while the broader coverage in "Solvents from waste" addresses the wider issue of waste valorization to make sustainable solvents.

When we consider wastes as feedstocks, it is important that we do not forget carbon dioxide. This major natural chemical that is a vital part of our life cycles and of the critical interaction between animal and plant life, has become regarded as a threat to civilization through its overproduction resulting from our uncontrolled burning of fossil fuels. From a biorefinery perspective, CO_2 is a potential C_1 feedstock, and a number of synthesis pathways have been developed to make compounds from it. In particular, organic carbonates can be synthesized using CO_2 and alcohols, making them potentially 100% bio-based, at least for those small alcohols that are currently made from biomass. The resulting carbonates are considerably more attractive, at least from an environmental perspective, than those made using phosgene. The use of organic carbonates as solvent is the subject of a chapter in this book.

Solvents continue to play a key role in almost every industry sector. In the last 50 or so years we have built up an impressive array of solvents that offer a remarkable diversity of properties to suit an equally diverse range of applications. The challenge for "green chemistry" is to find safe, sustainable and effective replacements so that we can continue to enjoy the benefits of solvents without the environmental harm. Bio-based solvents will play an essential role in this quest, and this book helps to show us how.

James Clark
University of York Green Chemistry Centre of Excellence
April 2017

1

Glycerol as Eco-Efficient Solvent for Organic Transformations

Palanisamy Ravichandiran and Yanlong Gu

School of Chemistry and Chemical Engineering, Huazhong University of Science and Technology, Wuhan, China

1.1 Introduction

Organic solvents are used in the chemical and pharmaceutical industries [1]. The global demand for these solvents has reached 20 million metric tons annually [2]. Solvents are unreactive supplementary fluids that can dissolve starting materials and facilitate product separation through recrystallization or chromatographic techniques. In a reaction mixture, the solvent is involved in intermolecular interactions and performs the following: (i) stabilization of solutes, (ii) promoting the preferred equilibrium position, (iii) changing the kinetic profile of the reaction, and (iv) influencing the product selectivity [3]. Selection of appropriate solvents for organic transformations is important to develop green synthesis pathways using renewable feedstock. In the past two decades, green methodologies and solvents have gained increasing attention because of their excellent physical and chemical properties [4–6]. Green solvents should be non-flammable, biodegradable and widely available from renewable sources [7].

Bio-Based Solvents, First Edition. Edited by François Jérôme and Rafael Luque.
© 2017 John Wiley & Sons Ltd. Published 2017 by John Wiley & Sons Ltd.

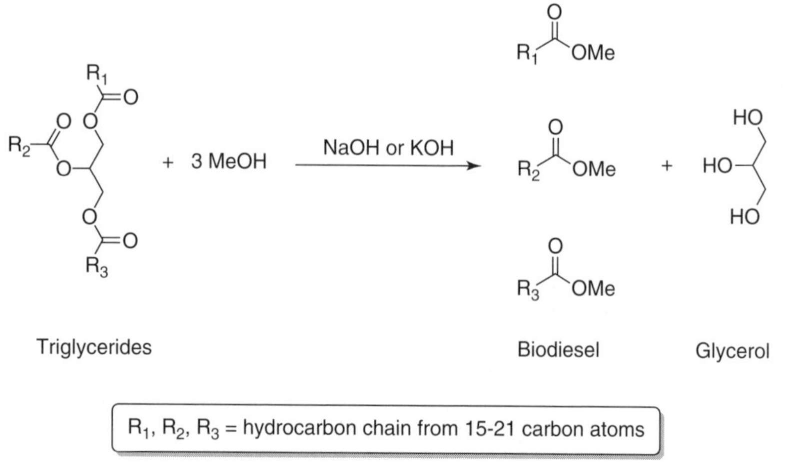

R₁, R₂, R₃ = hydrocarbon chain from 15-21 carbon atoms

Figure 1.1 Reaction for biodiesel production.

Biodiesel production involves simple catalytic transesterification of triglycerides under basic conditions (Figure 1.1) [8]. This process generates glycerol as a by-product (approximately 10% by weight). The amount of glycerol produced globally has reached 1.2 million tons and will continue to increase in the future because of increasing demand for biodiesel [9]. Glycerol has more than 2000 applications, and its derivatives are highly valued starting materials for the preparation of drugs, food, beverages, chemicals and synthetic materials (Figure 1.2) [10].

The biodiesel industries generate large amounts of glycerol as a by-product. As such, the price of glycerol is low, leading to its imbalanced supply. Currently, a significant proportion of this renewable chemical is wasted. This phenomenon has resulted in a negative feedback on the future economic viability of the biodiesel industry and adversely affects the environment because of improper disposal [11]. In this regard, the application of glycerol as a sustainable and green solvent has been investigated in a number of organic transformations (Table 1.1). Glycerol is a colourless, odourless, relatively safe, inexpensive, viscous, hydroscopic polyol, and a widely available green solvent. Glycerol acts as an active hydrogen donor in several organic reactions. Glycerol exhibits a high boiling point, polarity and non-flammability and is a suitable substitute for organic solvents, such as water, dimethylformamide (DMF) and dimethylsulfoxide (DMSO). Thus, glycerol is considered a green solvent and an important subject of research on green chemistry. This review provides new perspectives for minimizing glycerol wastes produced by biomass industries.

Our research group has contributed a comprehensive review on green and unconventional bio-based solvents for organic reactions [12]. However, enthusiasm for using glycerol as a green solvent for organic transformations in particular continues to increase. The present paper thus summarizes recent developments

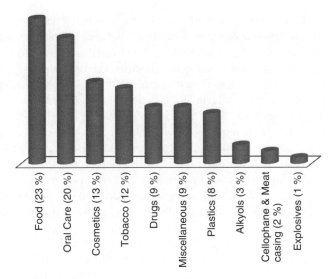

Figure 1.2 Commercial consumption of glycerol (industrial sectors and volumes). (A colour version of this figure appears in the plate section.)

Table 1.1 *Physical, chemical and toxicity properties of glycerol.*

Melting point	17.8°C
Boiling point	290°C
Viscosity (20°C)	1200 cP
Vapour pressure (20°C)	<1 mm Hg
Density (20°C)	1.26 g cm^{-3}
Flash point	160°C (closed cup)
Auto-ignition temperature	400°C
Critical temperature	492.2°C
Critical pressure	42.5 atm
Dielectric constant (25°C)	44.38
Dipole moment (30–50°C)	2.68 D
LD$_{50}$ (oral, rat)	12600 mg kg^{-1}
LD$_{50}$ (dermal, rabbit)	>10 000 mg kg^{-1}
LD$_{50}$ (rat, 1 h)	570 mg m^{-3}

on metal-free and metal-promoted organic reactions in glycerol between 2002 and 2016.

1.2   Metal-Free Organic Transformations in Glycerol

The synthesis of complex organic molecules utilizes harsh reaction conditions, expensive reagents and toxic organic solvents. Most organic transformations use expensive metal catalysts, such as Pd(OAc)$_2$, PdCl$_2$, PtCl$_2$ and AuCl$_2$.

The drawbacks of metal-promoted organic reactions are categorized into the following: (i) isolation and reuse of catalysts, (ii) lack of catalytic efficiency in the second usage, and (iii) disposal of metal catalysts. Over the past three decades, both industrial and academic chemists have continuously explored suitable methodologies, such as the use of green solvents. The chemical synthesis of glycerol as a sustainable solvent has gained wide attention because it provides valuable chemical scaffolds. Sugar fermentation produces glycerol either directly or as a by-product of the conversion of lignocelluloses into ethanol. Glycerol promotes this reaction without requiring any metal catalysts because of its excellent physical properties. Moreover, glycerol is widely available from renewable feedstock and is thus an appropriate green solvent for various reactions [13].

Quinoxaline, benzoxazole and benzimidazole derivatives can be synthesized using different methods; these molecules are commonly prepared through the condensation reaction of aryl 1,2-diamine with 1,2-dicarbonyl compounds [14, 15]. Bachhav *et al.* [16] developed an efficient, catalyst-free and straightforward method for synthesis of quinoxaline, benzoxazole and benzimidazole ring systems in glycerol; the yield is higher than those of conventional methods. The substrates 2-aminophenol and benzaldehyde are used as counter-reagents for the preparation of 2-arylbenzoxazoles (**1**). This reaction was tested in different solvent systems, but the desired products were not obtained and only glycerol efficiently promoted the reaction (Scheme 1.1). Radatz *et al.* [17] reported the same condensation reaction between 1,2-diamine and 1,2-dicarbonyl compound for the synthesis of benzodiazepines and benzimidazoles; the catalyst-free condensation reaction between *o*-phenylenediamine and benzaldehyde in glycerol produces benzimidazoles (**2**). The reaction between *o*-phenylenediamine and ketones produces benzodiazepines (**3**) when using glycerol as solvent. Furthermore, glycerol can be easily recovered and reused for condensation. However, at the fourth time of using glycerol, it starts to lose its activity (Scheme 1.2).

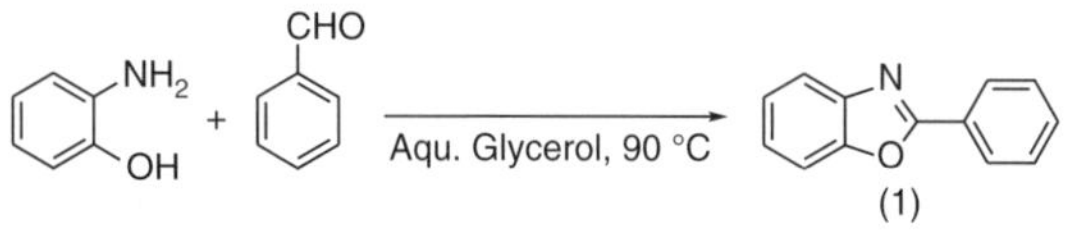

Scheme 1.1   Catalyst-free selective synthesis of 2-phenylbenzoxazole.

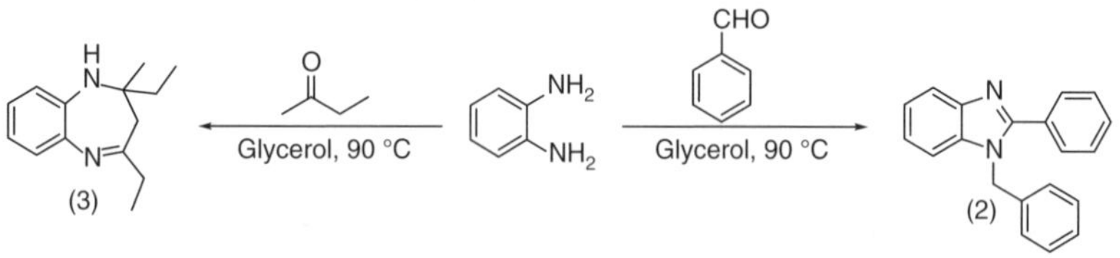

Scheme 1.2   Green synthesis of benzimidazoles and benzodiazepines in glycerol.

Nascimento *et al.* [18] used a similar kind of methodology for one-pot hetero-Diels–Alder reaction between (*R*)-citronellal and substituted arylamines; glycerol was used as a green, sustainable and recyclable solvent for the catalyst-free reaction and functioned as a model substrate to produce octahydroacridines (**4,5**) at high percentage yields with an isomeric product ratio of 46 : 54. The reactions proceeded without using acid catalysts (Scheme 1.3). Somwanshi *et al.* [19] developed a catalyst-free one-pot imino Diels–Alder reaction, with aldehydes, amines and cyclic enol ethers as model substrates to prepare the desired furano- and pyranoquinolines (**6**). 3,4-Dihydro-2*H*-pyran was used to prepare pyranoquinolines, and the mixture was obtained as endo-isomers, exo-isomers and errandendo-diastereomers; when furans were used as reagents, a single isomeric product is produced. Glycerol can be used as a sustainable solvent and leads to more efficient reactions than those using other organic solvents (Scheme 1.4).

Cabrera *et al.* [20] confirmed that glycerol is an efficient solvent for the oxidation of aromatic, aliphatic and functionalized thiols under microwave conditions. The oxidation reactions proceeded quickly, and the preferred disulfides (**7**) were obtained in good to excellent yields. Thiophenol, a strong nucleophile, was used as model substrate, and sodium carbonate, an inorganic base, was used as catalyst. Glycerol was easily recovered and used for further oxidation of thiols (Scheme 1.5). Zhou *et al.* [21] reported the condensation reaction between

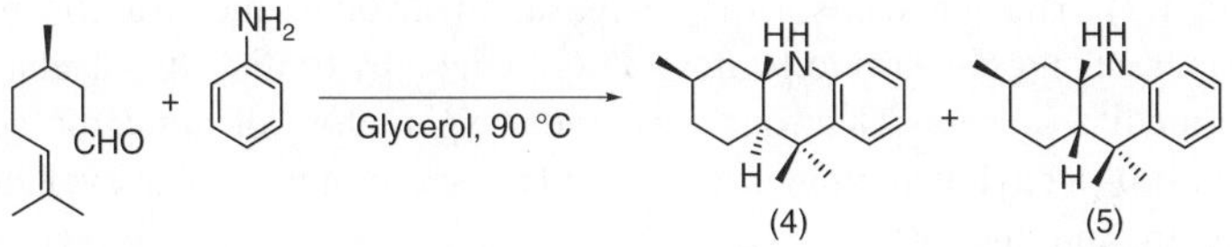

Scheme 1.3 Catalyst-free synthesis of isomeric mixtures of octahydroacridines.

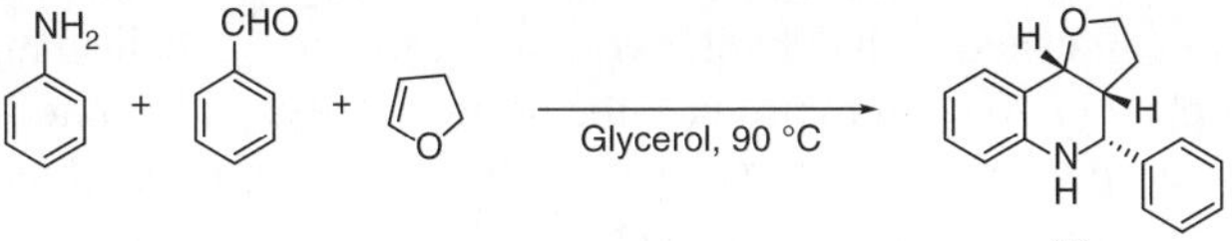

Scheme 1.4 One-pot synthesis of furanoquinolines by green method.

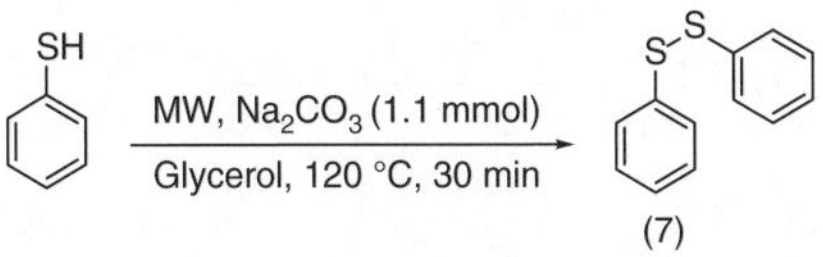

Scheme 1.5 Microwave-assisted synthesis of phenyl disulfide.

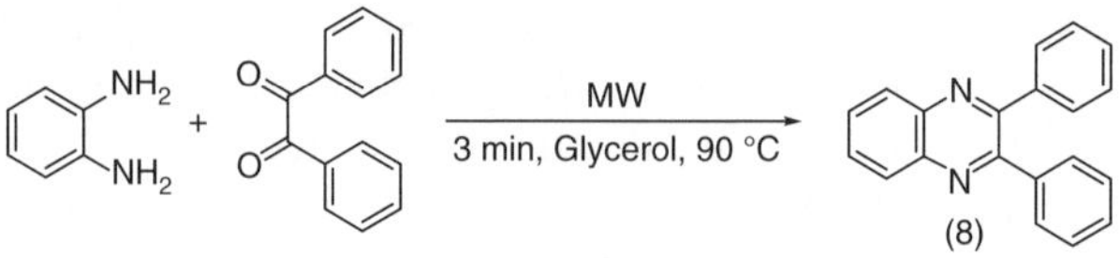

Scheme 1.6   Green synthesis of 2,3-diphenyl-quinoxaline in glycerol.

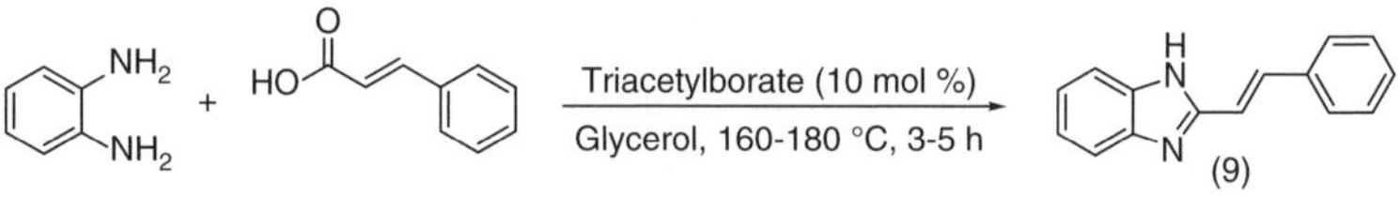

Scheme 1.7   Triacetylborate-catalysed green synthesis of 2-styryl-1*H*-benzimidazole.

1,2-diamines and 1,2-dicarbonyl compounds for synthesis of quinoxaline derivatives in glycerol at 90°C for 3 minutes without adding inorganic salts; the desired product (**8**) exhibited a high degree of purity (Scheme 1.6).

The synthesis of benzimidazole scaffolds has attracted wide interest because of their biological activities. Taduri *et al.* [22] performed a green synthesis of 2-heterostyrylbenzimidazole derivatives. The condensation reaction between *o*-phenylenediamine and 3-phenylacrylic acid in glycerol was catalysed by 10 mol.% triacetylborate and produced the desired bicyclic benzimidazoles (**9**) (Scheme 1.7). The mechanism of condensation involves the reaction between triacetylborate and glycerol to form an intermediate by trailing two acetate units; the intermediate accepts electrons from the oxygen atom of cinnamic acid and reacts with *o*-phenylenediamine, which undergoes intramolecular cyclization to form the final product (**8**).

N-Aryl phthalimide derivatives are important organic skeletons applied in biology and electronics because of their structural arrangements. Lobo *et al.* [23] reported that glycerol promoted the synthesis of *N*-aryl phthalimide; the model reaction was conducted with dehydrative condensation of phthalic anhydride and aryl nucleophile in glycerol to produce the desired *N*-aryl phthalimides (**10**) with high yield. Deep eutectic solvents, such as choline chloride : malonic acid/urea (1 : 1, v/v), were also efficient solvents and catalysts for this reaction system (Scheme 1.8).

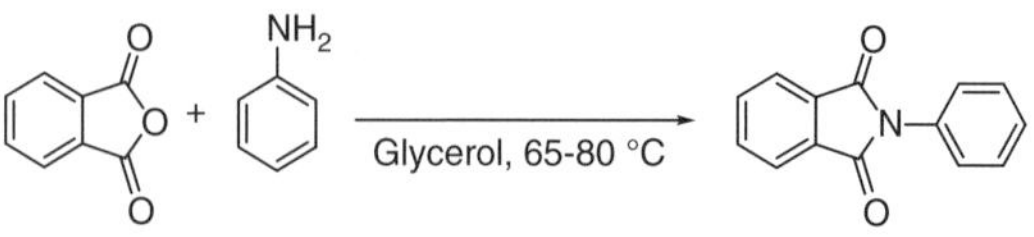

Scheme 1.8   Glycerol-mediated synthesis of *N*-aryl phthalimides.

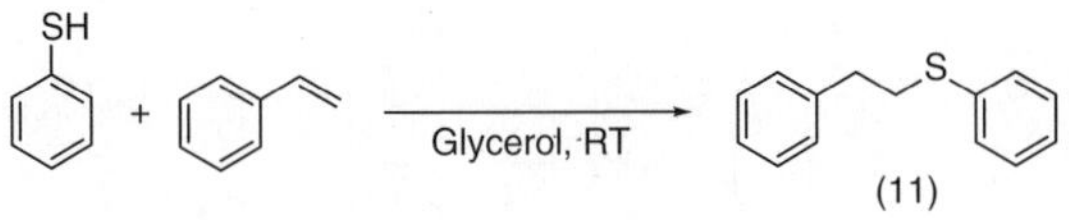

(11)

Scheme 1.9   Metal-free synthesis of thioethers at room temperature.

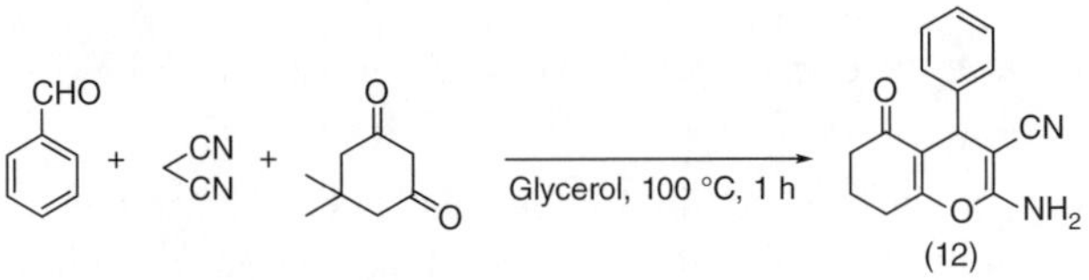

(12)

Scheme 1.10   Green synthesis of 4*H*-pyrans.

In the synthesis of linear thioethers, glycerol was used as an efficient and recyclable solvent for addition of thiols to non-activated alkenes [24]; the mock-up reaction was performed with styrene and thiophenol in glycerol at room temperature or under heating conditions to produce the desired thioethers (**11**) with 94% yield. Glycerol is considered a renewable and non-toxic solvent because of its excellent physical and chemical properties. This solvent can be utilized in green methods to effectively synthesize novel thioeugenols with antioxidant activities (Scheme 1.9).

Safaei *et al.* [25] used glycerol, a biodegradable and reusable medium, as solvent for one-pot three-component synthesis of 4*H*-pyrans under catalyst-free conditions, with benzaldehyde, dimedone and malononitrile as model substrates; the reaction follows the tandem Knoevenagel condensation and Michael-like addition in glycerol to produce the desired product 4*H*-pyrans (**12**). Solvents such as poly(ethylene glycol) 400 (PEG400), ethylene glycol and water also exhibited high product yields (83%, 80% and 71%, respectively) (Scheme 1.10). Shekouhy and co-workers [26] used the same reaction to synthesize simple urazole derivatives within a short reaction period; the model multi-component reaction utilized benzaldehyde condensed with *N*-phenyl urazole and an active methylene compound (dimedone) in glycerol to produce 94% of the desired urazole derivative (**13**). Glycerol plays an important role, that is, its polar amphoteric hydroxyl groups easily react with weakly acidic and basic components in the reaction because hydrogen bonds stabilize the intermediates (Scheme 1.11).

Thurow *et al.* [27] synthesized arylselanylanilines, which are important medicinal scaffolds and molecular skeletons. This synthesis was realized by the reaction between *N,N*-dimethylaniline and phenylselanyl chloride in glycerol, resulting in 99% yield of the desired product (**14**). In this Mannich-type reaction, *N,N*-dimethylaniline as nucleophile attacks phenylselanyl chloride at the *para*-position to form the intermediate aryliminium, which undergoes proton

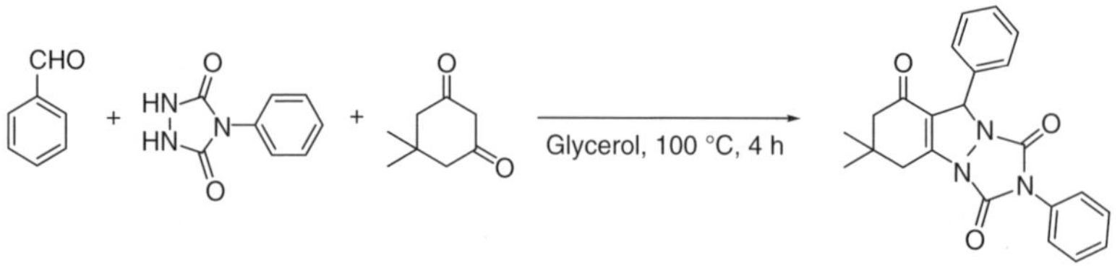

(13)

Scheme 1.11 Glycerol-promoted synthesis of triazolo[1,2-a]indazole-triones.

elimination to produce the desired compound (**14**). Glycerol plays a unique role in the reaction because it stabilizes the hydrogen bonds with the iminium intermediate (Scheme 1.12).

Vanillin semicarbazone derivatives are potent molecules for the treatment of antimicrobial diseases and are abundant in natural products [28, 29]. Jovanović *et al.* [30] synthesized vanillin and semicarbazone derivatives through green carbonyl–amine condensation; the reaction used vanillin and semicarbazone in glycerol solvent and was performed at 65°C for 20 minutes to produce the desired Schiff base (**15**). The same reaction conducted in ethanol provided high yield, that is, 72% of the desired vanillin semicarbazone, but required strong acid catalysts, such as sulfuric acid (Scheme 1.13).

A recent study elaborated a well-organized, green synthesis of 1-amidoalkyl-2-naphthols (**16**) through one-pot three-component condensation of aromatic aldehydes, acetamide and 2-naphthol, in glycerol solvent, with glycerosulfonic acid as catalyst (Scheme 1.14). The interaction of glycerosulfonic

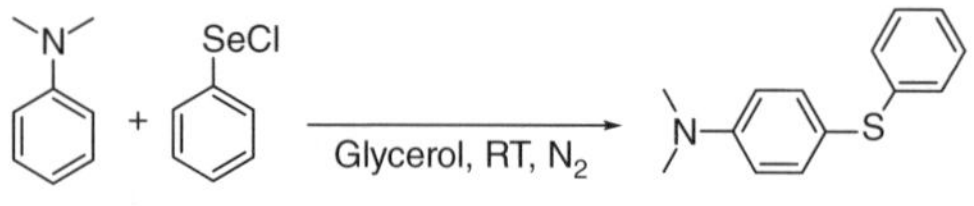

(14)

Scheme 1.12   Synthesis of dimethyl-(4-phenylsulfanyl-phenyl)-amine.

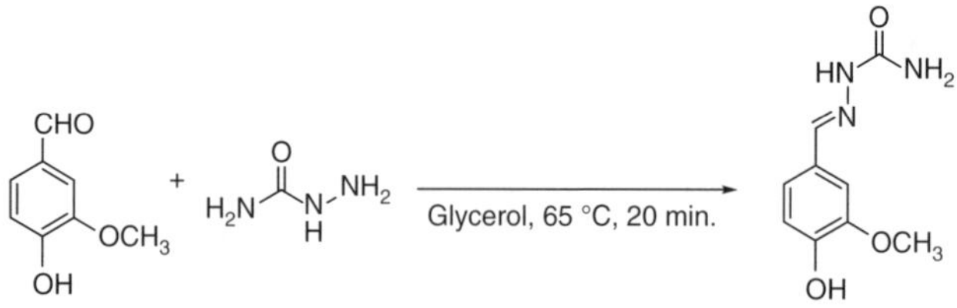

(15)

Scheme 1.13   Catalyst-free green protocol for synthesis of vanillin semicarbazone.

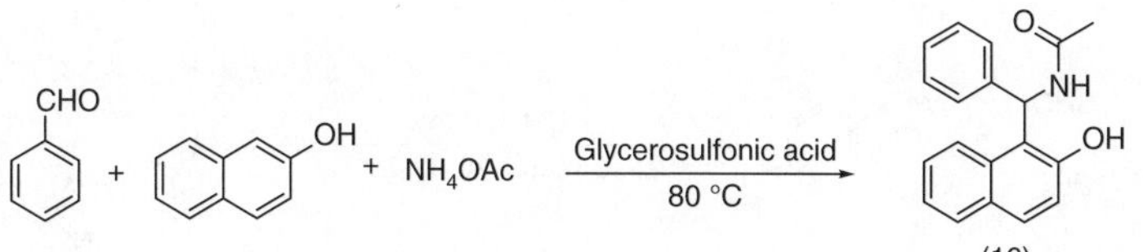

Scheme 1.14   Glycerosulfonic acid-promoted synthesis of 1-amidoalkyl-2-naphthol.

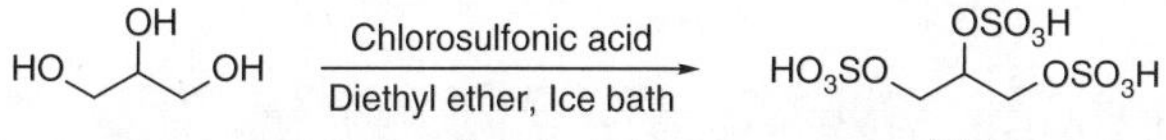

Scheme 1.15   Preparation of glycerosulfonic acid.

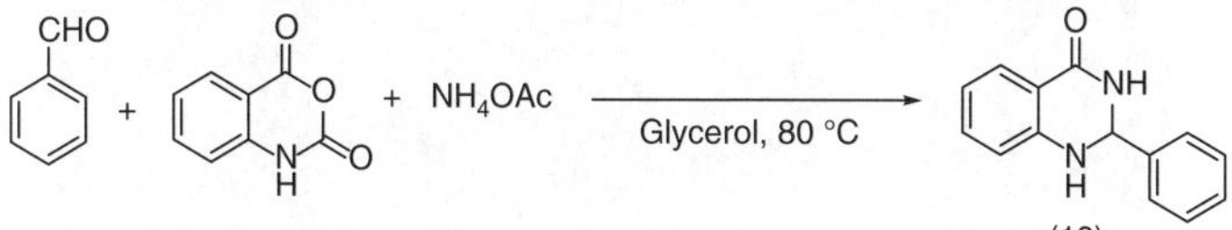

Scheme 1.16   Synthesis of 2,3-dihydroquinazoline-4(1*H*)-one.

acid with chlorosulfonic acid in diethyl ether produced compound (**17**) (Scheme 1.15). A simple method for one-pot three-component synthesis of 2,3-dihydroquinazoline-4(1*H*)-ones (**18**) has also been established using isatoic anhydride, aldehydes and ammonium acetate as model substrates in glycerol as green catalyst and solvent (Scheme 1.16). Thus, glycerol can be used as an effective and valuable solvent for synthesis of 2,3-dihydroquinazoline-4(1*H*)-ones and 1-amidoalkyl-2-naphthols [31].

He *et al.* [32] reported the glycerol-promoted synthesis of di(indolyl)methanes, xanthene-1,8(2*H*)-diones and 1-oxo-hexahydroxanthenes through catalyst-free electrophilic activation of aldehydes. The model reaction was carried out with 4-nitrobenzaldehyde and 2-methylindole in glycerol solvent to produce the desired compound, namely, di(indolyl)methanes (**19**). The glycerol-mediated reactions obtained 95% yield, and 2.0 ml of glycerol was required to proceed the reaction (Scheme 1.17). Glycerol is used as solvent in organic reactions because its hydrophilic nature facilitates the isolation of the product. The same protocol was used to isolate the product by adding water containing glycerol to the reaction mixture; the product settled down in water and was separated by vacuum filtration (Figure 1.3).

Nascimento *et al.* [33] developed a simple direct cyclocondensation reaction with model substrates α-arylselanyl-1,3-diketones and arylhydrazines, which

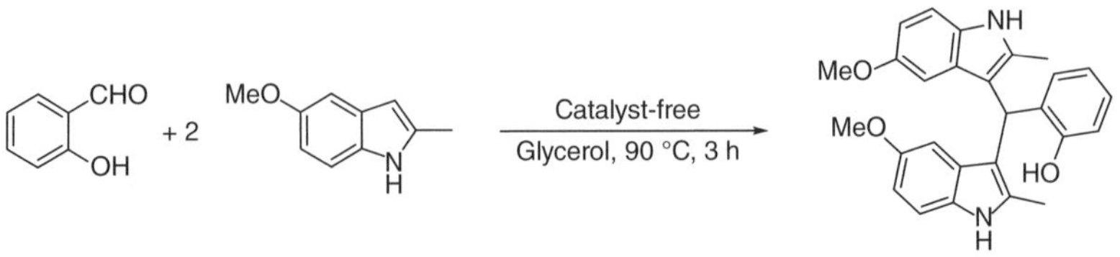

(19)

Scheme 1.17 Catalyst-free green synthesis of di(indolyl)methane.

(a)   (b)   (c)

Figure 1.3 Development of the model reaction in glycerol [32]: (a) beginning of the reaction as an identical mixture; (b) partial precipitation of the reaction; (c) the end of the precipitation of the reaction. From He *et al.* (2009) *Green Chem.*, **11**, 1767–1773. Reproduced by permission of RSC. (A colour version of this figure appears in the plate section.)

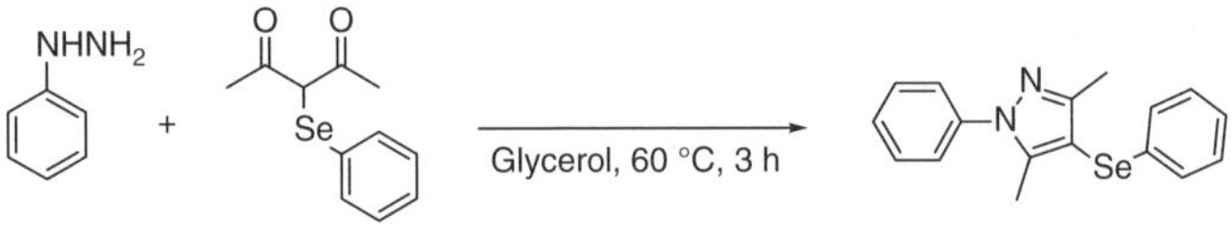

(20)

Scheme 1.18 Metal-free synthesis of 4-arylselanylpyrazoles.

possess both electron-withdrawing and electron-donating groups, to obtain high yields of 4-arylselanylpyrazoles (**20**) (Scheme 1.18). Similarly, Min *et al.* [34] demonstrated the same cyclocondensation reaction, which can be promoted using an economical and environment-friendly solvent, namely, glycerol–water system. Phenylhydrazine and 1,3-dicarbonyl compounds were used as benchmark partners to prepare the desired pyrazoles (**21**) (Scheme 1.19).

Rodriguez *et al.* [35] developed an intermolecular azide–alkyne Huisgen cycloaddition reaction between diphenylacetylene and benzylazide to produce

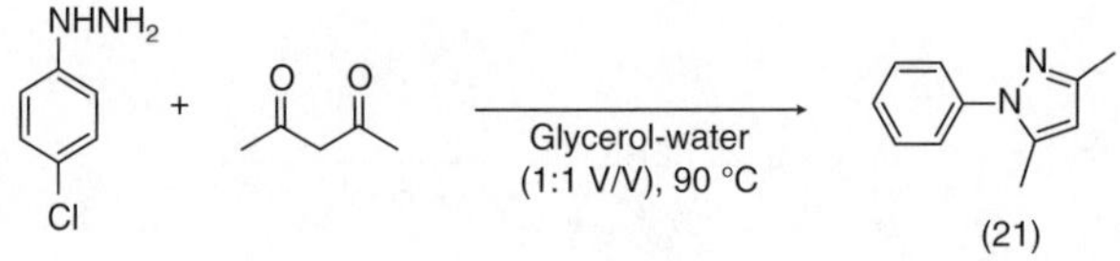

(21)

Scheme 1.19 Direct cyclocondensation reaction between phenylhydrazine and pentane-2,4-dione in glycerol–water system.

high yields of 1,2,3-triazole (**22**). Non-activated internal alkynes have been converted in neat glycerol under thermal and microwave dielectric heating. The important role of glycerol in the reaction has been confirmed by theoretical density functional theory (DFT) calculations. The DFT results revealed that the BnN_3/glycerol adduct promotes the reaction and stabilizes the corresponding lowest unoccupied molecular orbital (LUMO), thereby increasing the reactivity with alkyne in the first case (Scheme 1.20).

The multi-component Petasis borono–Mannich (PBM) reaction is a useful tool for preparing complex tertiary morpholine derivatives (**23**). Boronic acids, aldehydes/ketones and amines can be used as promoters in the reaction. Glycerol provided high product yield, and crude glycerol was considered an appropriate solvent for successful PBM reaction (Scheme 1.21) [36].

Ganesan *et al.* [37] developed a three-component Betti reaction with non-toxic and inexpensive glycerol as solvent. A reaction without catalyst was also performed, and the desired Betti bases (**24**) were isolated, providing up to 91% yield. The reaction in glycerol works well for all kinds of amines and aldehydes. The reaction in 20 mol.% methanesulfonic acid as catalyst also produced 93% benzoxanthene (Scheme 1.22).

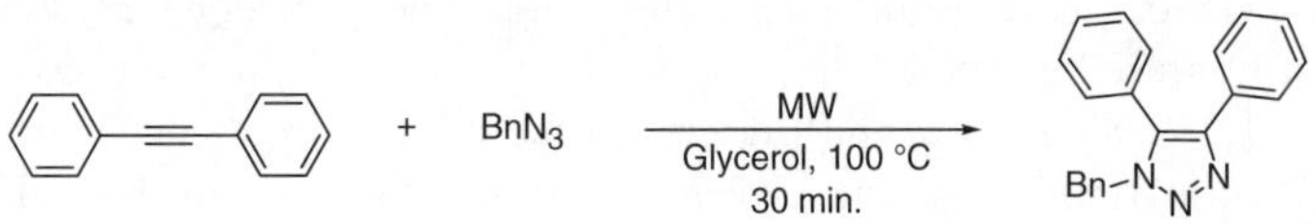

(22)

Scheme 1.20 Huisgen cycloaddition reaction for preparing 1,2,3-triazoles.

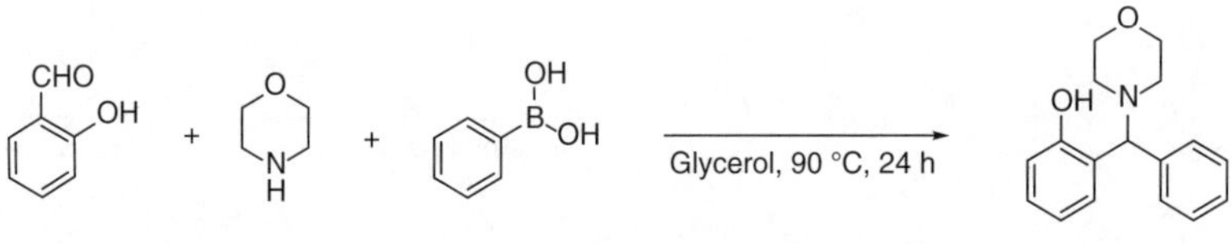

(23)

Scheme 1.21 PBM reaction for preparing complex tertiary morpholine derivatives.

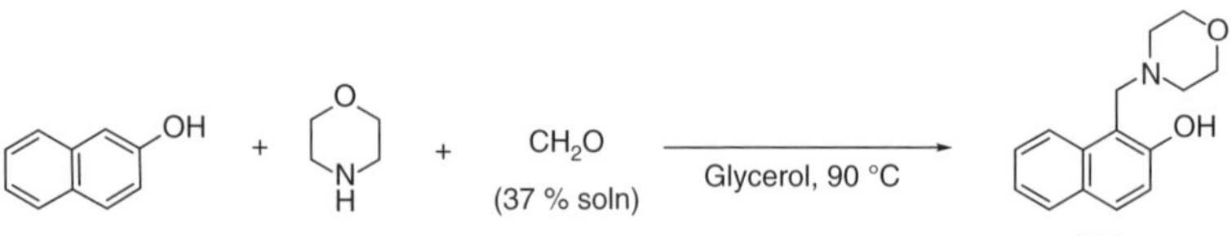

(24)

Scheme 1.22   Preparation of 1-(morpholinomethyl)naphthalen-2-ol through the Betti reaction.

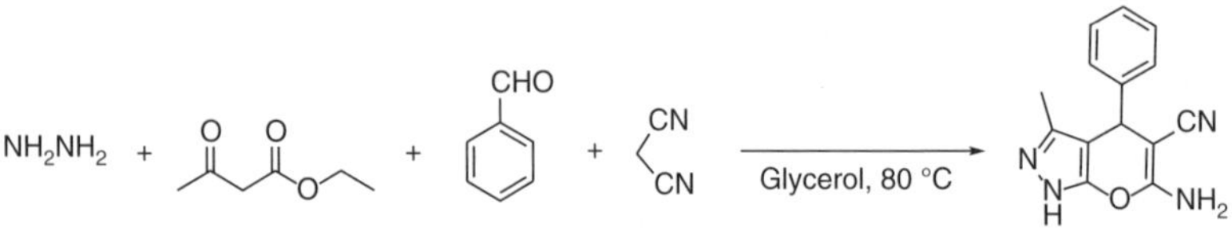

(25)

Scheme 1.23   Glycerol-mediated, one-pot synthesis of dihydropyrano[2,3-*c*]pyrazoles.

Sohal *et al.* [38] proposed a multi-component, one-pot synthesis of various pyrazole derivatives through the condensation of ethyl acetoacetate, hydrazine, aromatic aldehyde and malononitrile, with glycerol as solvent. The reaction involved the condensation between hydrazine and ethylacetoacetate; the Knoevenagel condensation occurred between malononitrile and aldehyde to form arylidenepropanedinitrile as intermediate; and the Michael-like addition to produce the desired pyrazole derivative (**25**) at high yield (Scheme 1.23).

Singh *et al.* [39] developed the first glycerol-promoted green synthesis of spirooxindole-indazolones and spirooxindole-pyrazolines; this multi-component tandem reaction used isatin, phenyl hydrazine and dimedone as substrates. Glycerol was used as solvent to promote the formation of the dimedone anion (Scheme 1.24). Singh *et al.* [40] described the catalyst-free facile synthesis of pyrido[2,3-*d*]pyrimidines using glycerol as promoting medium. Benzaldehyde, malononitrile and uracil used as substrates were condensed through sequential reaction, involving Knoevenagel condensation, Michael-type addition and air

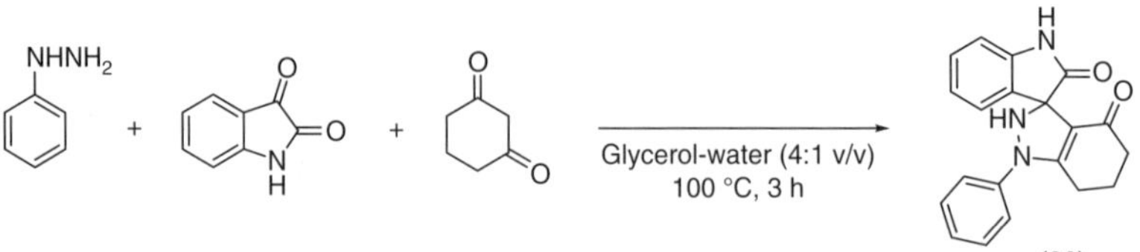

(26)

Scheme 1.24   Catalyst-free synthesis of spirooxindole-indazolones.

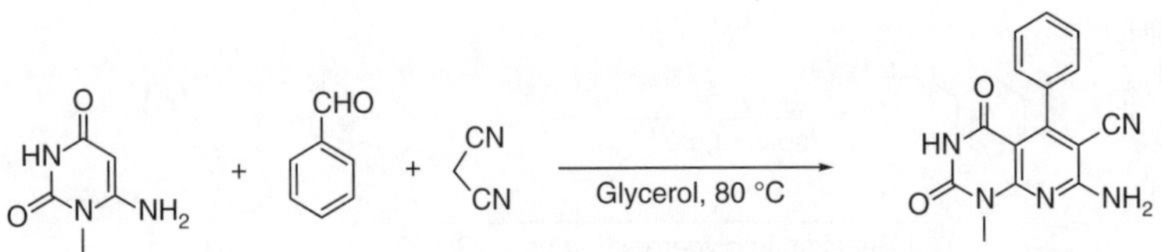

Scheme 1.25   Multi-component tandem synthesis of pyrido[2,3-*d*]pyrimidine.

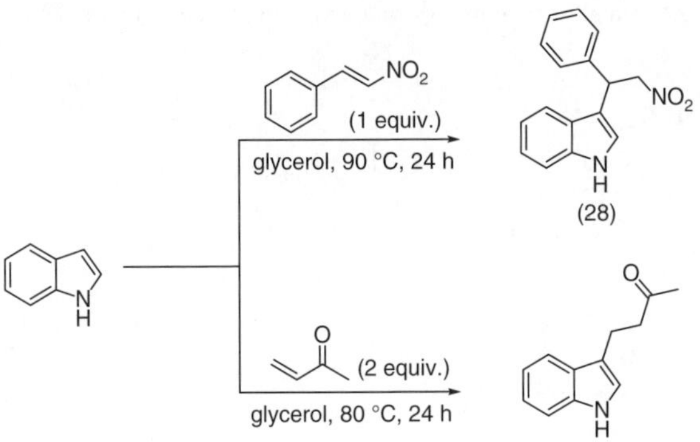

Scheme 1.26   Friedel–Crafts alkylation of indoles in glycerol.

oxidation, to produce a high yield of the desired pyrido[2,3-*d*]pyrimidine (**27**) (Scheme 1.25).

For the Michael-type reaction of indole to the α,β-unsaturated compound, conventional organic solvents, such as DMSO, toluene, 1,2-dichloroethane or DMF, are ineffective under catalyst-free conditions. Low amounts of products (**28,29**) were formed in the Michael addition of indoles to β-nitrostyrene and 3-buten-2-one in water. However, the reaction in glycerol produced high yields of the desired products (**28,29**) without using any Brønsted or Lewis acid catalysts (Scheme 1.26) [41].

Glycerol is also an exceptional supporting medium for the ring-opening reaction of styrene oxide with *p*-anisidine. The reaction did not use any catalysts but showed good selectivity. In aqueous medium the selectivity of the reaction products **30 : 31** was in the ratio 76 : 24, and that in glycerol was 93 : 7 under identical conditions (Scheme 1.27) [42].

Tan *et al.* [43] developed a coupling reaction of phenylhydrazine, β-keto ester, formaldehyde and 1-ethyl-2-phenylindole in glycerol solvent to form a highly functionalized indole derivative (**32**; 42% yield) (Scheme 1.28).

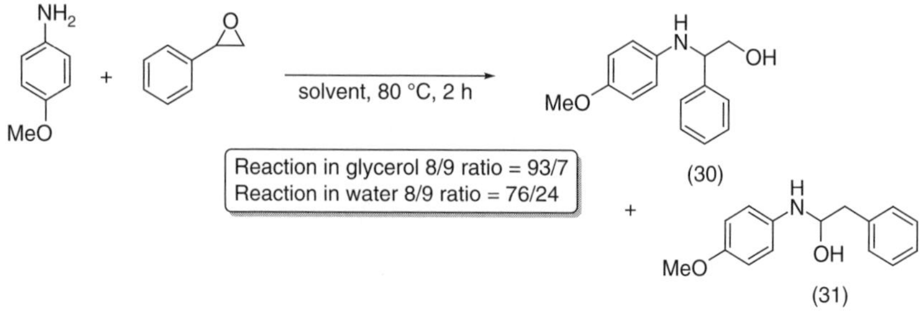

Scheme 1.27 Catalyst-free ring-opening reaction of styrene oxide with *p*-anisidine.

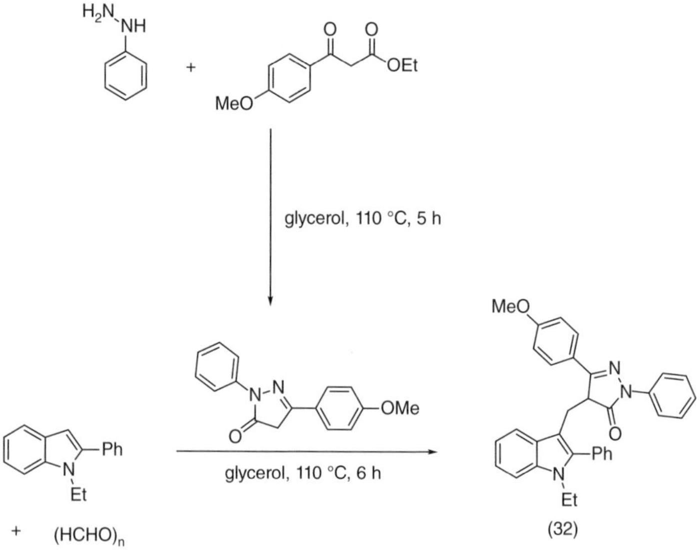

Scheme 1.28 Coupling reaction for synthesis of the functionalized indoles.

The multi-component reactions with three or more reactants are combined in a single-step reaction to facilitate diverse complex molecules. Therefore, glycerol was established to be a convenient medium for multi-component reactions. Styrene, primary amines, 2-naphthol, 4-hydroxy-6-methyl-2-pyrone and 4-hydroxy-1-methyl-2-quinolone are readily available molecules for use as target partners accumulated with 1,3-cyclohexadiones and formaldehyde in glycerol to obtain complex polycyclic compounds (**33–37**) under catalyst-free conditions (Scheme 1.29). In these multi-component reactions, glycerol not only provided efficient reaction but also allowed simple separation of the products (**33–37**) through extraction because of its strong hydrophilic nature [44].

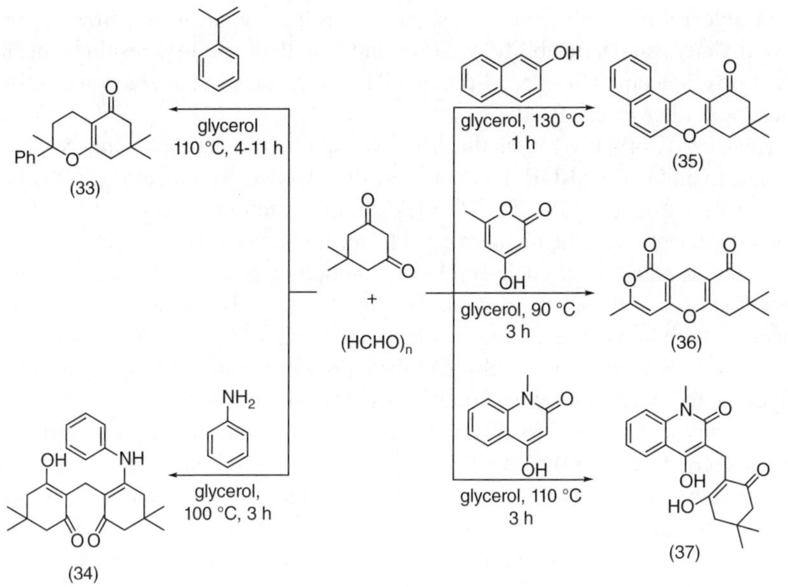

Scheme 1.29 Multi-component reactions of 1,3-cyclohexanediones and formaldehyde in glycerol.

1.3 Metal-Promoted Organic Transformations in Glycerol

Metal-free organic reactions in glycerol are commonly used because of their green properties, despite the fact that complex molecular skeletons can be synthesized through metal-catalysed organic reactions. Catalysis plays a prominent role in this type of green organic synthesis because it allows economical and environment-friendly efficient preparation of chemicals and materials [45]. New selective catalysts can provide short-cuts in the total synthesis but remain inefficient. Currently, metal-promoted green organic synthesis is attracting considerable attention and is documented through published books [46] and articles [47]. Over the past decades, chemical industries have applied a number of homogeneous catalysts to prepare bulk chemicals. The important and versatile property of these homogeneous catalysts is attributed to the tunability of transition metal complexes by changing the ligands coordinated with the metal [48]. Homogeneous catalysts cannot be reused because they are difficult to isolate from the reaction mixture. By contrast, heterogeneous catalysts can be easily separated from the reaction and are thus commonly used in organic preparations. Metal catalysts embedded with nanoparticles can be efficient catalysts for a number of organic transformations. These metal catalysts in green solvents, such as water, glycerol, poly(ethylene glycol) and gluconic acid, also provide interesting medicinal scaffolds. In particular, metal-catalysed organic reactions with glycerol as solvent medium show interesting chemical properties, such as low toxicity, biodegradability, polarity and non-flammability, resulting in high reaction efficiency. Thus, metal-promoted green synthesis with glycerol as solvent must be further investigated.

Khatri *et al.* [49] used aryl halides in glycerol for *N*-arylation of amines. Copper acetate (5 mol.%) and KOH (2 mmol) were dissolved in glycerol at 100°C to obtain the desired product (**38**) with 96% yield. In this reaction, glycerol acts as ligand and coordinates with the metal catalyst to accelerate the reaction (Scheme 1.30).

Lenardão *et al.* [50] used glycerol as a promoting medium for the cross-coupling reaction of diaryldiselenides with (*Z*) or (*E*) vinyl bromides, which contain electron-withdrawing and electron-donating groups. The mock cross-coupling reaction was carried out with (*E*)-β-bromostyrene and diphenyldiselenide in glycerol and catalysed by the combined system of copper iodide CuI (5 mol.%) and Zn dust (0.6 mmol) at 110°C in the absence of oxygen atmosphere to produce the desired product (**39**) with 98% yield (Scheme 1.31). Lenardão *et al.* [51] also reported the synthesis of 2-organylselanyl pyridines. The reaction was promoted by glycerol under nitrogen atmosphere. Interestingly, this reaction did not use any metal catalysts for cross-coupling of diphenyldiselenide. The reaction was carried out with diphenyldiselenide and 2-chloropyridinein glycerol with H_3PO_4 (1.0 ml) as reducing agent to produce the targeted product, 2-phenylselanyl-pyridine (**40**), with 99% yield (Scheme 1.32). The same research group [52] reported that PEG400 and glycerol promoted the green synthesis of organylthioenynes at high yields and high selectivity with KF/Al_2O_3 as catalytic system. The catalytic mixture KF/Al_2O_3 (0.7 g) was used to promote the reaction and produce the *Z* and *E* mixture of organylthioenynes (**41,42**) (Scheme 1.33). The same group used a green solvent and catalytic system for one-pot synthesis of β-aryl-β-sulfanyl

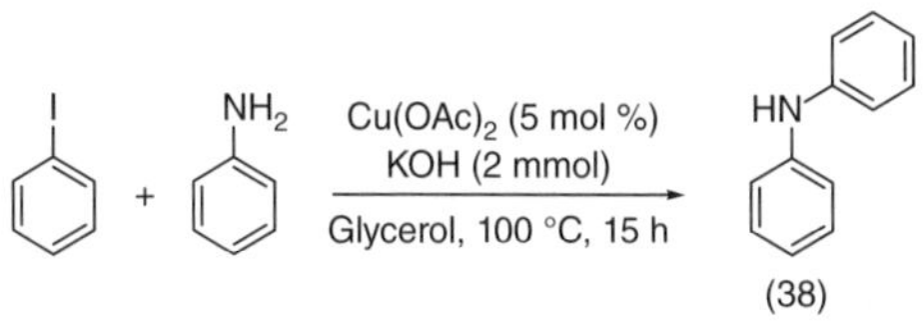

Scheme 1.30   Copper acetate-promoted *N*-arylation of aromatic amine.

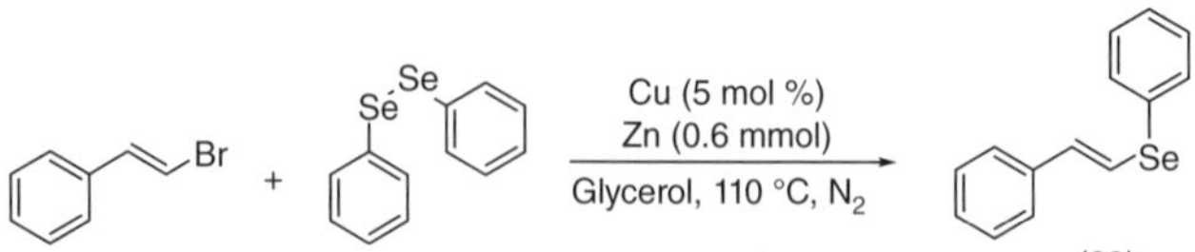

Scheme 1.31   CuI-promoted green synthesis of phenyl selenide derivatives.

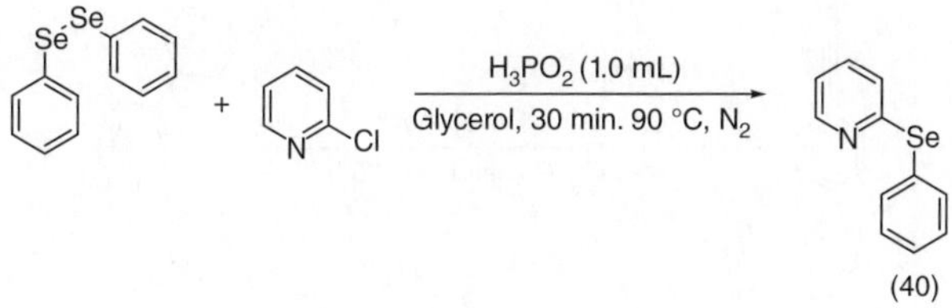

(40)

Scheme 1.32   Synthesis of 2-phenylselanyl-pyridine.

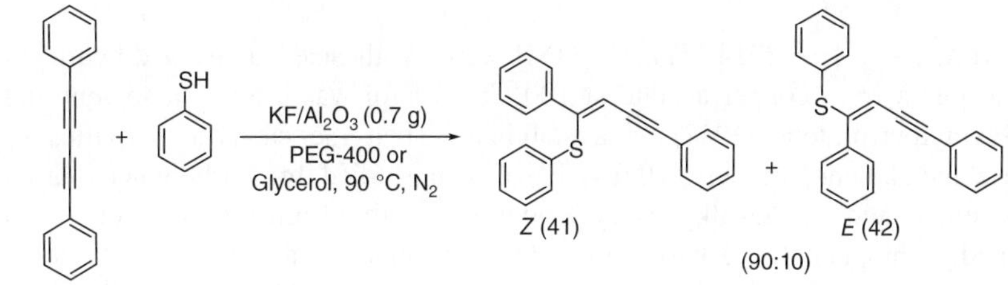

Z (41)   E (42)

(90:10)

Scheme 1.33   Green solvent and catalyst promoted the synthesis of a Z and E mixture of organylthioenynes.

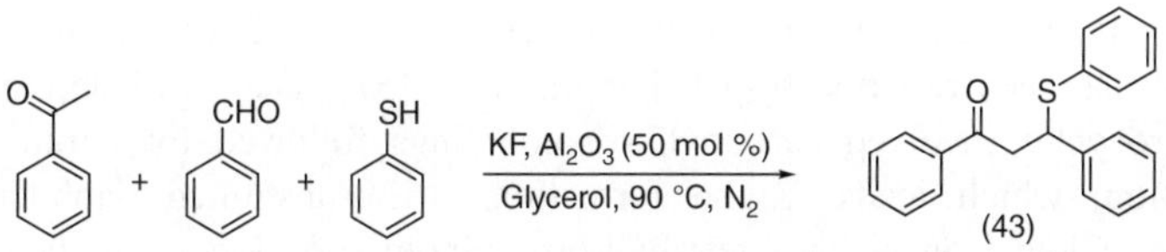

(43)

Scheme 1.34   Green synthesis of β-aryl-β-sulfanyl ketones.

ketones [53]. One-pot solid-supported green synthesis was carried out with acetophenone, active carbonyl compounds and nucleophile in glycerol. Notably, 50 mol.% KF/Al_2O_3 was needed to produce the desired β-aryl-β-sulfanyl ketones (**43**) (Scheme 1.34).

The mixture of CuI and glycerol displayed versatile catalytic activity in the Huisgen cycloaddition of azides and terminal or 1-iodoalkynes [54]. The base-free Cu(I)-catalysed 1,3-dipolar cycloaddition of azides with terminal and 1-iodoalkynes were synthesized and reported by Vidal *et al.* [55] in 2014. The reaction started with benzyl azide and phenylacetylene in pure glycerol; it was catalysed by 1 mol.% of CuI to produce an enormous 1,3-dipolar cycloaddition product (**44**) with 99% yield (Scheme 1.35). In order to prove the efficiency of glycerol in the reaction, other solvents were also used to promote the reaction; however, they produced only marginal product yield.

Cross-coupling and azide–alkyne cycloaddition processes have been catalysed by copper oxide nanoparticles (Cu_2ONP), as reported by Chahdoura and

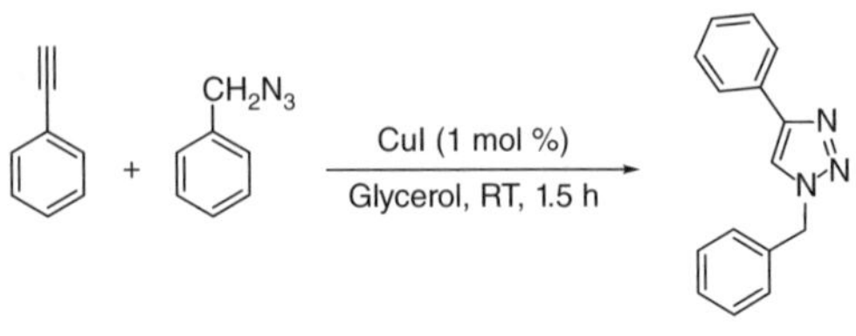

(44)

Scheme 1.35   Copper(I)-catalysed synthesis of 1-benzyl-4-phenyl-1*H*-[1–3]triazole.

co-workers [56] in 2014. The Cu_2ONP were synthesized under a dihydrogen atmosphere from copper acetate at 100°C; glycerol was used as a solvent and polyvinylpyrrolidone (PVP) as a stabilizer. Then the catalytic activities of these metal nanoparticles (MNPs) have been studied for carbon–heteroatom couplings and azide–alkyne cycloaddition with 4-nitro-iodobenzene and 4-methyl-thiophenolin, a base, *tert*-BuOK, and glycerol at 100°C for 24 hours to produce the desired polyfunctional product (**45**) with 89% yield without the isolation of intermediates. The scope of the catalyst and substrates were further extended and a library of organic scaffolds has been synthesized and reported (Scheme 1.36).

Catalysis by palladium nanoparticles (PdNP) in a green synthesis using glycerol has been reported by Chahdoura *et al.* [57]. The multi-step synthesis of heterocycles involved carbonylative couplings followed by intramolecular cyclization, which leads to the formation of *N*-substituted naphthalimides (**46**), isoindole-1-ones and tetrahydroisoquinolin-1,3-diones in better isolated yields. Apart from this synthesis, the PdNP were also employed to synthesize 2-benzofurans and dihydrobenzofurans via Sonogashira coupling/heterocyclisation tandem processes (Scheme 1.37).

Recently, the cross-coupling reactions were tested again for the catalytic system bis(2-pyridyl)diselenoethers [58]. Coupling reagents, such as 2-iodo-1,3-dimethoxy-benzene, and the strong counter-nucleophile, thiophenol, were used as model substrates, glycerol as a greener solvent, and an inorganic salt, KOH (3 equiv), as a base. The ligand, $[Cu_4I_4\{(2\text{-PySe})_2CH_2\}_2]$, was used as the metal ligand in the reaction, and the desired product 1,3-dimethoxy-2-phenylsulfanyl-benzene (**47**) was isolated from the reaction with 63% yield. Other ligands, such as $[CuCl_2\{(2\text{-PySe})_2CH_2\}]_n$,

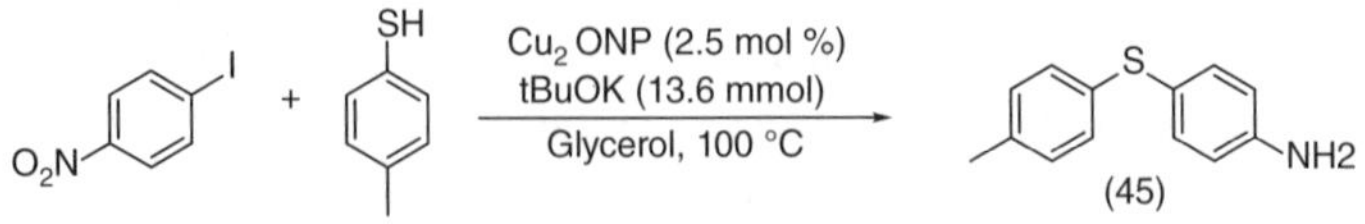

(45)

Scheme 1.36   Copper nanoparticle-promoted synthesis of 4-*p*-tolylsulfanyl-phenylamine.

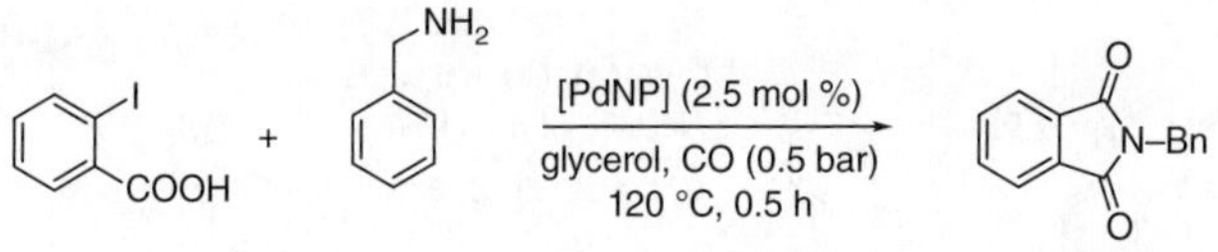

(46)

Scheme 1.37 Palladium nanoparticle-catalysed synthesis of *N*-substituted phthalimides.

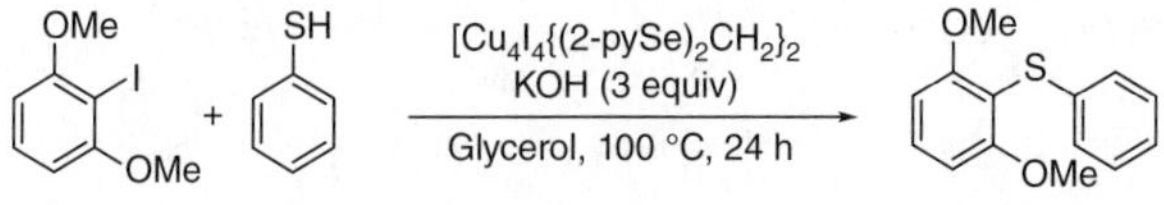

(47)

Scheme 1.38 Metal ligand-promoted green synthesis of bis(2-pyridyl)diselenoethers.

$[CuCl_2\{(2-PySe)_2(CH_2)_3\}_2]$, were also used in the reaction to prepare the desired C–S coupling products (Scheme 1.38).

Glycerol is a sustainable green solvent for many organic transformations. Nevertheless, it has drawbacks, such as low solubility of gases, high viscosity and high hydrophobicity. These problems can be overcome by using some other new techniques in a stand-alone manner, such as microwaves (MW) [59] or high-intensity ultrasound (US) [60, 61], or in a combined manner [62, 63], which results in the enhancement of the reaction rates. A similar kind of reaction protocol has been demonstrated by Cravotto *et al.* [64]. They realized a transfer hydrogenation reaction of benzaldehyde using glycerol as dual solvent and hydrogen donor. By using Ru(*p*-cumene)Cl$_2$ as catalyst, the reaction proceeded with 100% conversion in the presence of bases (NaOH + KOH) under ultrasonic conditions, to give the desired product, phenyl-methanol (**48**) (Scheme 1.39).

The cross-coupling reactions for the construction of C–C and C–heteroatom bonds have attracted considerable attention because of their significance in medicinal chemistry [65, 66]. The formation of one C–N bond and one C–C bond on the identical carbenic centre has been developed and reported by Aziz *et al.* [67]. A hypothesis for this reaction was that it involves a copper acetate-catalysed cross-coupling reaction between 2′-bromo-biaryl-*N*-tosylhydrazones and different

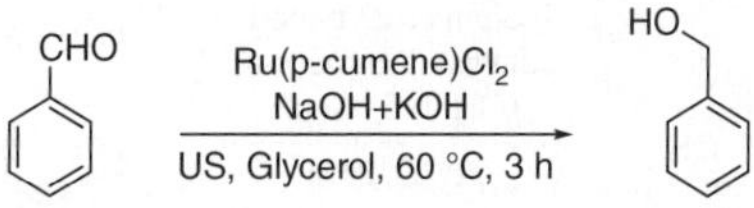

(48)

Scheme 1.39 Ultrasound-promoted green synthesis of phenyl-methanol.

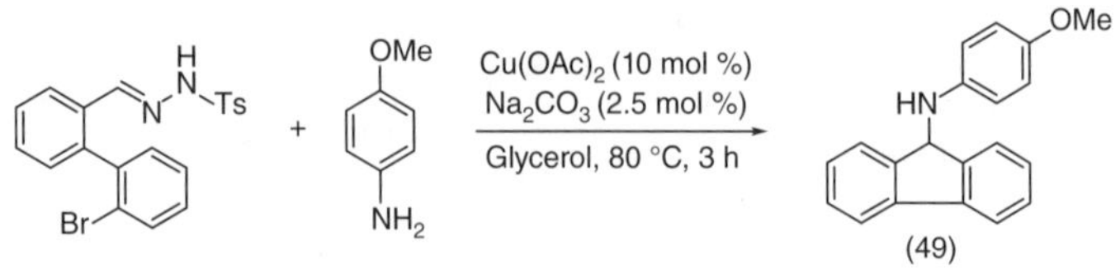

Scheme 1.40 Copper-catalysed cross-coupling between *N*-tosylhydrazones and 4-methoxyaniline.

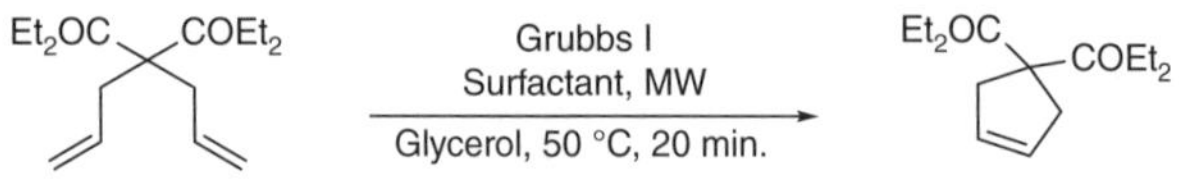

Scheme 1.41 Microwave-assisted ring-closing metathesis of diethyl diallylmalonate.

amines as benchmark partners, leading to the formation of 9*H*-fluoren-9-amine derivatives (**49**). This reaction proceeds under mild conditions with 2.5 mol.% sodium carbonate in glycerol, a cheap and environmentally gracious solvent, without external ligand (Scheme 1.40).

The microwave-assisted ring-closing metathesis has also been studied recently by Hamel and co-workers [68]. The reaction was started with diethyl diallylmalonate in glycerol under microwave irradiation. Catalysis with Grubbs I catalyst generates the desired five-membered cyclic product (**50**). The reaction proceeding without the catalyst generates by-products (Scheme 1.41).

Selective hydrogen transfer methodologies for the reduction of ketones or imines and oxidation of alcohols and amines, and the donation of protons (typically 2-propanol) or electrons (e.g. acetone) are easily conducted in green solvents like water or glycerol [69, 70]. The selective transfer of hydrogen from organic carbonyl compounds has also been studied recently by Azua *et al.* [71]. Ir(III) complexes, which possess chelating bis-NHC ligand and sulfonate groups, are efficient catalysts because of their excellent solubility in the reaction medium and the strong electron-donor nature of the bis-carbene ligands. KOH was used to accelerate the reaction and reduce acetophenone in glycerol to produce the desired hydrogen transferred product, 1-phenyl-ethanol (**51**), and by-product, 1,3-dihydroxy-propan-2-one (**52**) (Scheme 1.42).

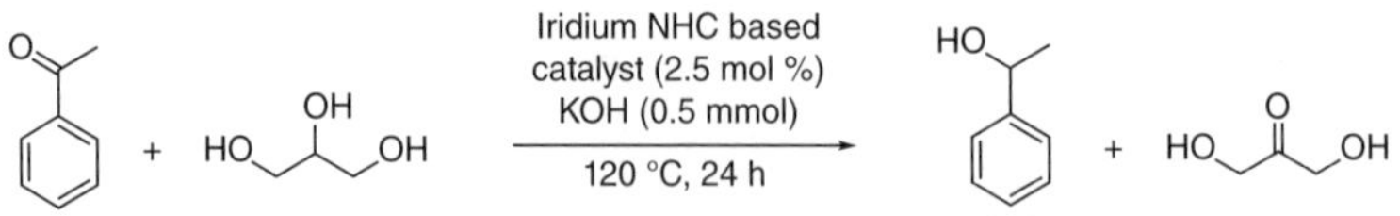

Scheme 1.42 Iridium-catalysed selective hydrogen transfer of acetophenone in glycerol.

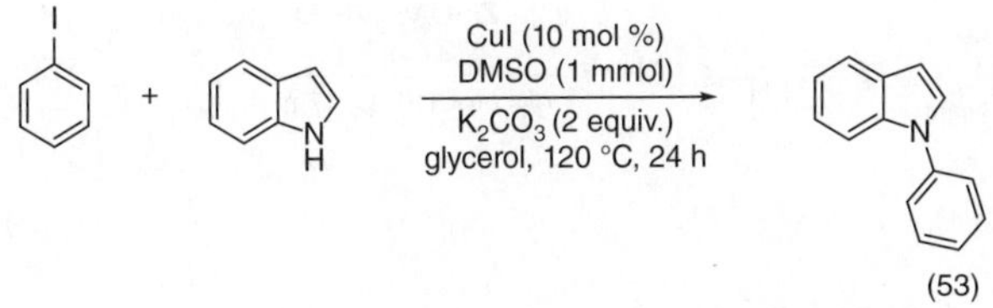

Scheme 1.43 Copper/glycerol catalytic system for the synthesis of *N*-aryl indoles.

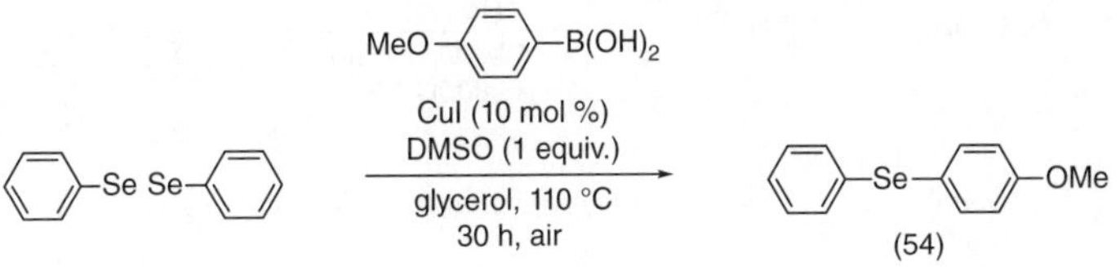

Scheme 1.44 Synthesis of diarylselenides using glycerol as solvent.

The copper-catalysed synthesis of different *N*-aryl indoles (**53**) has been developed by Yadav *et al.* [72]. The cross-coupling reaction of indoles with aryl halides used glycerol as a green sustainable solvent and DMSO as an additive to accelerate the reaction. The low catalytic amount of CuI (10 mol.%) in the reaction provided various *N*-aryl indoles in good to outstanding yield (Scheme 1.43). Similarly, the copper-catalysed cross-coupling reactions of diaryldiselenides with aryl boronic acids have been reported by Ricordi *et al.* [73]; glycerol promoted cross-coupling to produce the corresponding diarylselenides (**54**) (Scheme 1.44).

The reactions at the carbon–carbon double bond of (*E*)-chalcones have been described by Mesquita *et al.* [74]. The reaction of diphenyldiselenide in glycerol at 90°C with H_3PO_2 as a reducing agent with different (*E*)-chalcones produced chemoselective 1,4-reduction products (**55**). Moreover, under similar conditions, the natural product, zingerone, was synthesized and obtained in high yield (Scheme 1.45).

The combined catalytic system of glycerol and zinc(II) acetate for the synthesis of 2-pyridyl-2-oxazolines (**56**) was developed by Carmona *et al.* [75]. The reaction

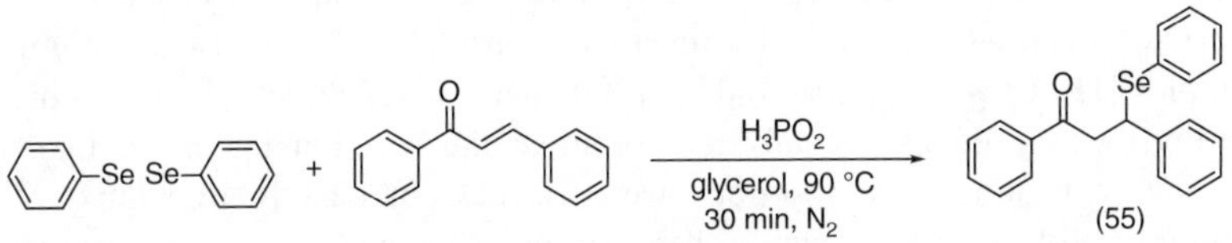

Scheme 1.45 Synthesis of 1,3-diphenyl-3-(phenylselanyl)propan-1-one under a green protocol.

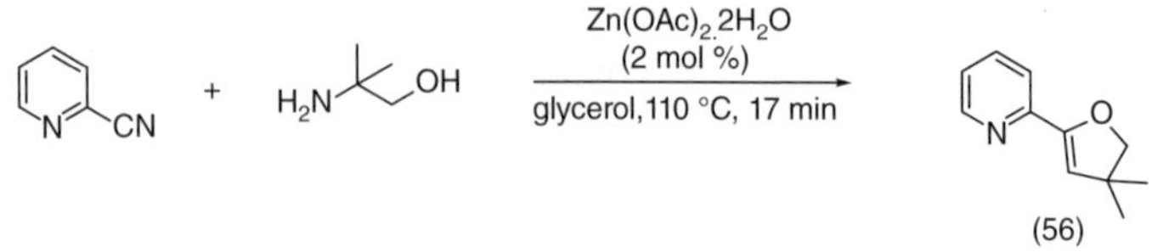

(56)

Scheme 1.46   Oxazoline synthesis using glycerol as solvent.

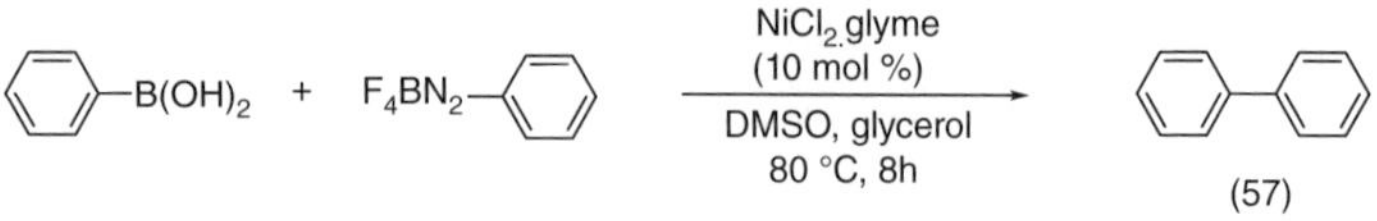

(57)

Scheme 1.47   Suzuki coupling of phenyl boronic acids in glycerol.

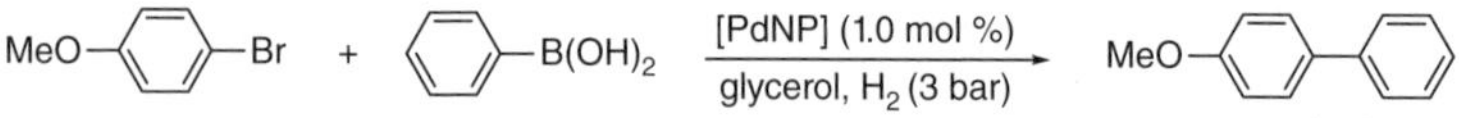

(58)

Scheme 1.48   Suzuki–Miyaura coupling reaction of aryl halides in glycerol.

started with amino alcohols and 2-cyanopyridines under microwave irradiation and resulted in the formation of the desired 2-pyridyl-2-oxazoline derivatives. Conventional heating did not significantly improve the reaction, and the catalytic system was recovered and reused for further synthesis (Scheme 1.46).

The base-free nickel-catalysed Suzuki coupling reaction of phenyl boronic acids with aryl diazonium salts in glycerol has been described by Bhojane *et al.* [76]. Different aryl diazonium salts reacted with aryl boronic acids under the optimized conditions to obtain high yields of the corresponding diaryl compounds (**57**) (Scheme 1.47). Similarly, a Suzuki–Miyaura coupling reaction of aryl halides has been demonstrated by Chahdoura *et al.* [77]. Palladium nanoparticles catalysed the reaction efficiently to produce the desired coupling product (**58**). An interesting catalytic immobilization in glycerol was observed for C–C coupling (up to 10 runs without loss of any catalytic activity) (Scheme 1.48).

The aldoxime rearrangement to primary amides has been studied by González-Liste *et al.* [78]. Commercially available bis(allyl)-ruthenium(IV) complex [{RuCl(μ-Cl)(η^3:η^3-C$_{10}$H$_{16}$)}$_2$] (C$_{10}$H$_{16}$ = 2,7-dimethylocta-2,6-diene-1,8-diyl) was used as a catalyst to promote the reaction under thermal and microwave heating. The reactions were carried out cleanly in a mixture of water/glycerol (1 : 1). Aromatic, heteroaromatic, aliphatic and α,β-unsaturated aldoximes served as benchmark partners to produce the desired amides (**59**) in moderate to high yields (Scheme 1.49).

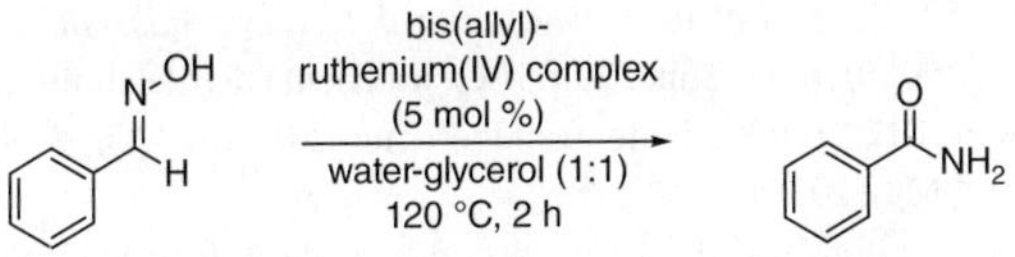

Scheme 1.49   Rearrangement of (*E*)-benzaldoxime into benzamide in glycerol.

1.4   Conclusions and Perspectives

We have described the beginning and development of the uses of glycerol as a green solvent, reagent and hydrogen donor for synthetic organic chemistry to transform the waste by-product into a useful solvent and products. Glycerol is very useful for organic reactions because of its availability from renewable sources on a bulk scale at cheap price and its advantageous properties, such as non-flammability, non-toxicity and biodegradability. Apart from the organic transformations, a number of nanocatalysts and room-temperature ionic liquids have also been prepared from glycerol, as described in this review. Use of glycerol as a solvent has several advantages and very few disadvantages. One of the latter is its viscosity, which could mean poor substrate contact with the solvent. The extraction of highly functionalized reaction products from glycerol is, in addition, a difficulty that needs to be solved prior to practical uses. Nevertheless, the above few problems will be overcome by the wise use of glycerol as a solvent. Glycerol has attracted considerable interest among green and organic chemists, and the authors believe that this inclusive literature will direct the further development in the field of glycerol chemistry.

Acknowledgements

The authors thank the National Natural Science Foundation of China for financial support (21173089 and 21373093). The authors are also grateful for Ms. Ping Liang and all staff members in the Analytical and Testing Center of HUST for their support and constant contributions to our work. The Cooperative Innovation Center of Hubei Province is also acknowledged. This work was also supported by the Fundamental Research Funds for the Central Universities in China (2014ZZGH019).

References

1. Ashcroft, C.P., Dunn, P., Hayler, J. and Wells, A.S. (2015) Survey of solvent usage in papers published in *Organic Process Research & Development* 1997–2012. *Org. Process Res. Dev.*, **19**, 740–747.

2. DECHEMA (2012) *Press Release: New natural resource base in the chemical industry – only a matter of time*, June. http://www.achema.de/fileadmin/user_upload/Bilder/Presse/ACHEMA2012/Trendberichte/Trendberichte_2012/tb19_en_Biobased_Chemicals .docx (accessed 4 May 2016).

3. Reichardt, C. and Welton, T. (2010) *Solvents and Solvent Effects in Organic Chemistry*, 4th edn, Wiley-VCH, Weinheim.

4. Handy, S.T. (2003) Greener solvents: room temperature ionic liquids from biorenewable sources. *Chem. Eur. J.*, **9**, 2938–2944.

5. Leitner, W. (2007) Editorial: Green solvents for processes. *Green Chem.*, **9**, 923.

6. Horváth, I.T. (2008) Solvents from nature. *Green Chem.*, **10**, 1024–1028.

7. Nelson, W.M. (2003) *Green Solvents for Chemistry: Perspectives and Practice*, Oxford University Press, Oxford.

8. Kemp, W.H. (2006) *Biodiesel: Basics and Beyond*, Azlext Press, Ontario.

9. Zhou, C.-H., Beltramini, J.N., Fan, Y.-X. and Lu, C.Q. (2008) Chemoselective catalytic conversion of glycerol as a biorenewable source to valuable commodity chemicals. *Chem. Soc. Rev.*, **37**, 527–549.

10. Gu, Y. and Jérôme, F. (2010) Glycerol as a sustainable solvent for green chemistry. *Green Chem.*, **12**, 1127–1138.

11. Pagliaro, M. and Rossi, M. (2008) *The Future of Glycerol: New Usages for a Versatile Raw Material*, RSC Publishing, Cambridge.

12. Gu, Y. and Jérôme, F. (2013) Bio-based solvents: an emerging generation of fluids for the design of eco-efficient processes in catalysis and organic chemistry. *Chem. Soc. Rev.*, **42**, 9550–9570.

13. Díaz-Álvarez, A.E., Francos, J., Lastra-Barreira, B., et al. (2011) Glycerol and derived solvents: new sustainable reaction media for organic synthesis. *Chem. Commun.*, **47**, 6208–6227.

14. Liu, C., Zhou, L., Jiang, D. and Gu, Y. (2016) Multicomponent reactions of aldo-X bifunctional reagent α-oxoketene dithioacetals and indoles or amines: divergent synthesis of dihydrocoumarins, quinolines, furans, and pyrroles. *Asian J. Org. Chem.*, **5**, 367–372.

15. Ravichandiran, P., Lai, B. and Gu, Y. (2017) Aldo-X bifunctional building blocks for the synthesis of heterocycles. *Chem. Rec.*, **17**, 142–183.

16. Bachhav, H.M., Bhagat, S.B. and Telvekar, V.N. (2011) Efficient protocol for the synthesis of quinoxaline, benzoxazole and benzimidazole derivatives using glycerol as green solvent. *Tetrahedron Lett.*, **52**, 5697–5701.

17. Radatz, C.S., Silva, R.B., Perin, G., et al. (2011) Catalyst-free synthesis of benzodiazepines and benzimidazoles using glycerol as recyclable solvent. *Tetrahedron Lett.*, **52**, 4132–4136.

18. Nascimento, J.E.R., Barcellos, A.M., Sachini, M., et al. (2011) Catalyst-free synthesis of octahydroacridines using glycerol as recyclable solvent. *Tetrahedron Lett.*, **52**, 2571–2574.

19. Somwanshi, J.L., Shinde, N.D. and Faruqui, M. (2013) Catalyst-free synthesis of furano- and pyranoquinolines by using glycerol as recyclable solvent. *Heterocyclic Lett.*, **3**, 69–74.

20. Cabrera, D.M.L., Líbero, F.M., Alves, D., et al. (2012) Glycerol as a recyclable solvent in a microwave-assisted synthesis of disulfides. *Green Chem. Lett. Rev.*, **5**, 329–336.

21. Zhou, W.J., Zhang, X.Z., Sun, X.B., et al. (2013) Microwave-assisted synthesis of quinoxaline derivatives using glycerol as a green solvent. *Russ. Chem. Bull.*, **62**, 1244–1247.

22. Taduri, A.K., Babu, P.N.K. and Devi, B.R. (2014) Glycerol containing triacetylborate mediated syntheses of novel 2-heterostyryl benzimidazole derivatives: a green approach. *Org. Chem. Int.*, **2014**, 260726.

23. Lobo, H.R., Singh, B.S. and Shankarling, G.S. (2012) Deep eutectic solvents and glycerol: a simple, environmentally benign and efficient catalyst/reaction media for synthesis of *N*-aryl phthalimide derivatives. *Green Chem. Lett. Rev.*, **5**, 487–533.

24. Lenardão, E.J., Jacob, R.G., Mesquita, K.D., et al. (2013) Glycerol as a promoting and recyclable medium for catalyst-free synthesis of linear thioethers: new antioxidants from eugenol. *Green Chem. Lett. Rev.*, **6**, 269–276.

25. Safaei, H.R., Shekouhy, M., Rahmanpur, S. and Shirinfeshan, A. (2012) Glycerol as a biodegradable and reusable promoting medium for the catalyst-free one-pot three component synthesis of 4*H*-pyrans. *Green Chem.*, **14**, 1696–1704.

26. Shekouhy, M., Sarvestani, A.M., Khajeh, S. and Nezhad, A.K. (2015) Glycerol: a more benign and biodegradable promoting medium for catalyst-free one-pot multi-component synthesis of triazolo[1,2-*a*]indazole-triones. *RSC Adv.*, **5**, 63705–63710.

27. Thurow, S., Penteado, F., Perin, G., et al. (2014) Metal and base-free synthesis of arylselanyl anilines using glycerol as a solvent. *Green Chem.*, **16**, 3854–3859.

28. Binil, P.S., Anoop, M.R., Jisha, K.R., et al. (2013) Growth, spectral and thermal characterization of vanillin semicarbazone (VNSC) single crystals. *J. Therm. Anal. Calorim.*, **111**, 575–581.

29. Ali, S.M.M., Azad, M.A.K., Jesmin, M., et al. (2012) *Asian Pac. J. Trop. Biomed.*, **2**, 438–442.

30. Jovanović, M.B., Konstantinović, S.S., Ilić, S.B. and Veljković, V.B. (2013) The synthesis of vanillin-semicarbazone in a crude glycerol as a green solvent. *Adv. Technol.*, **2**, 38–44.

31. Hajjami, M., Bakhti, F. and Ghiasbeygi, E. (2015) Incredible role of glycerol in multicomponent synthesis of 2,3-dihydroquinazoline-4(1*H*)-ones and 1-amidoalkyl-2-naphthols. *Croat. Chem. Acta*, **88**, 197–205.

32. He, F., Li, P., Gu, Y. and Li, G. (2009) Glycerol as a promoting medium for electrophilic activation of aldehydes: catalyst-free synthesis of di(indolyl)methanes, xanthene-1,8(2*H*)-diones and 1-oxo-hexahydroxanthenes. *Green Chem.*, **11**, 1767–1773.

33. Nascimento, J.E.R., de Oliveira, D.H., Abib, P.B., et al. (2015) Synthesis of 4-arylselanylpyrazoles through cyclocondensation reaction using glycerol as solvent. *J. Braz. Chem. Soc.*, **26**, 1533–1541.

34. Min, Z.-L., Zhang, Q., Hong, X., et al. (2015) A green protocol for catalyst-free syntheses of pyrazole in glycerol–water solution. *Asian J. Chem.*, **27**, 3205–3207.

35. Rodriguez-Rodriguez, M., Gras, E., Pericas, M.A. and Gomez, M. (2015) Metal-free intermolecular azide–alkyne cycloaddition promoted by glycerol. *Chemistry Eur. J.*, **21**, 18706–18710.

36. Rosholm, T., Gois, P.M.P., Franzen, R. and Candeias, N.R. (2015) Glycerol as an efficient medium for the Petasis borono–Mannich reaction. *ChemistryOPEN*, **4**, 39–46.

37. Ganesan, S.S., Rajendran, N., Sundarakumar, S.I., et al. (2013) β-Naphthol in glycerol: a versatile pair for efficient and convenient synthesis of aminonaphthols, naphtho-1,3-oxazines, and benzoxanthenes. *Synthesis*, **45**, 1564–1568.

38. Sohal, H.S., Goyal, A., Sharma, R., et al. (2013) Glycerol mediated, one pot, multicomponent synthesis of dihydropyrano[2,3-*c*]pyrazoles. *Eur. J. Chem.*, **4**, 450–453.

39. Singh, S., Saquib, M., Singh, S.B., et al. (2015) Catalyst free, multicomponent-tandem synthesis of spirooxindole-indazolones and spirooxindole-pyrazolines: a glycerol mediated green approach. *RSC Adv.*, **5**, 45152–45157.

40. Singh, S., Saquib, M., Singh, M., et al. (2016) A catalyst free, multicomponent-tandem, facile synthesis of pyrido[2,3-*d*]pyrimidines using glycerol as a recyclable promoting medium. *New J. Chem.*, **40**, 63–67.

41. Habib, P.M., Kavala, V., Kuo, C.-W. and Yao, C.-F. (2008) Catalyst-free aqueous-mediated conjugative addition of indoles to β-nitrostyrenes. *Tetrahedron Lett.*, **49**, 7005–7007.

42. Gu, Y., Barrault, J. and Jerome, F. (2008) Glycerol as an efficient promoting medium for organic reactions. *Adv. Synth. Catal.*, **350**, 2007–2012.

43. Tan, J.-N., Li, M. and Gu, Y. (2010) Multicomponent reactions of 1,3-disubstituted 5-pyrazolones and formaldehyde in environmentally benign solvent systems and their variations with more fundamental substrates. *Green Chem.*, **12**, 908–914.

44. Li, M., Chen, C., He, F. and Gu, Y. (2010) Multicomponent reactions of 1,3-cyclohexanediones and formaldehyde in glycerol: stabilization of paraformaldehyde in glycerol resulted from using dimedone as substrate. *Adv. Synth. Catal.*, **352**, 519–530.

45. Johansson Seechurn, C.C.C., Kitching, M.O., Colacot, T.J. and Snieckus, V. (2012) Palladium-catalyzed cross-coupling: a historical contextual perspective to the 2010 Nobel Prize. *Angew. Chem., Int. Edn*, **51**, 5062–5085.

46. Rylander, P.N. (1973) *Organic Syntheses with Noble Metal Catalysts*, Academic Press, New York.

47. Perin, G., Lenardão, E. J., Jacob, R.G. and Panatieri, R.B. (2009) Synthesis of vinyl selenides. *Chem. Rev.*, **109**, 1277–1301.

48. Leenders, S.H.A.M., Gramage-Doria, R., de Bruin, B. and Reek, J.N.H. (2015) Transition metal catalysis in confined spaces. *Chem. Soc. Rev.*, **44**, 433–448.

49. Khatri, P.K. and Jain, S.L. (2013) Glycerol ingrained copper: an efficient recyclable catalyst for the *N*-arylation of amines with aryl halides. *Tetrahedron Lett.*, **54**, 2740–2743.

50. Gonçalves, L.C., Fiss, G.F., Perin, G., et al. (2010) Glycerol as a promoting medium for cross-coupling reactions of diaryl diselenides with vinyl bromides. *Tetrahedron Lett.*, **51**, 6772–6775.

51. Thurow, S., Webber, R., Perin, G., et al. (2013) Glycerol/hypophosphorous acid: an efficient system solvent-reducing agent for the synthesis of 2-organylselanyl pyridines. *Tetrahedron Lett.*, **54**, 3215–3218.

52. Alves, D., Sachini, M., Jacob, R.G., et al. (2011) Synthesis of (Z)-organylthioenynes using KF/Al_2O_3/solvent as recyclable system. *Tetrahedron Lett.*, **52**, 133–135.

53. Perin, G., Mesquita, K., Calheiro, T.P., et al. (2014) Synthesis of β-aryl-β-sulfanyl ketones by a sequential one-pot reaction using KF/Al_2O_3 in glycerol. *Synth. Commun.*, **44**, 49–58.

54. Díaz-Álvarez, E., Francos, J., Crochet, P. and Cadierno, V. (2014) Recent advances in the use of glycerol as green solvent for synthetic organic chemistry. *Curr. Green Chem.*, **1**, 51–65.

55. Vidal, C. and García-Álvarez, J. (2014) Glycerol: a biorenewable solvent for base-free Cu(I)-catalyzed 1,3-dipolar cycloaddition of azides with terminal and 1-iodoalkynes. Highly efficient transformations and catalyst recycling. *Green Chem.*, **16**, 3515–3521.

56. Chahdoura, F., Pradel, C. and Gómez, M. (2014) Copper(I) oxide nanoparticles in glycerol: a convenient catalyst for cross-coupling and azide–alkyne cycloaddition processes. *ChemCatChem*, **6**, 2929–2936.

57. Chahdoura, F., Mallet-Ladeira, S. and Gómez, M. (2015) Palladium nanoparticles in glycerol: a clear-cut catalyst for one-pot multi-step processes applied in the synthesis of heterocyclic compounds. *Org. Chem. Front.*, **2**, 312–318.

58. Cargnelutti, R., da Silva, F.D., Abram, U. and Lang, E.S. (2015) Metal complexes with bis(2-pyridyl)diselenoethers: structural chemistry and catalysis. *New J. Chem.*, **59**, 7948–7953.

59. Kappe, C.O. and Stadler, A. (2005) *Microwaves in Organic and Medicinal Chemistry*, Wiley-VCH, Weinheim.

60. Cravotto, G., Nano, G.M., Palmisano, G. and Tagliapietra, S. (2003) The reactivity of 4-hydroxycoumarin under heterogeneous high-intensity sonochemical conditions. *Synthesis*, **8**, 1286–1291.

61. Cravotto, G. and Cintas, P. (2006) Power ultrasound in organic synthesis: moving cavitational chemistry from academia to innovative and large-scale applications. *Chem. Soc. Rev.*, **35**, 180–196.

62. Cravotto, G., Garella, D., Calcio Gaudino, E. and Leveque, J.-M. (2008) Microwaves–ultrasound coupling: a tool for process intensification in organic synthesis. *Chim. Oggi*, **26** (2), 44–46.

63. Cravotto, G. and Cintas, P. (2007) The combined use of microwaves and ultrasound: improved tools in process chemistry and organic synthesis. *Chemistry, Eur. J.*, **13**, 1902–1909.

64. Cravotto, G., Orio, L., Calcio Gaudino, E., et al. (2011) Efficient synthetic protocols in glycerol under heterogeneous catalysis. *ChemSusChem*, **4**, 1130–1134.

65. Barluenga, J. and Valdés, C. (2011) Tosylhydrazones: new uses for classic reagents in palladium-catalyzed cross-coupling and metal-free reactions. *Angew. Chem., Int. Edn*, **50**, 7486–7500.

66. Fulton, J.R., Aggarwal, V.K. and de Vicente, J. (2005) The use of tosylhydrazone salts as a safe alternative for handling diazo compounds and their applications in organic synthesis. *Eur. J. Org. Chem.*, **2005**, 1479–1492.

67. Aziz, J., Frison, G., Gómez, M., et al. (2014) Copper-catalyzed coupling of *N*-tosylhydrazones with amines: synthesis of fluorene derivatives. *ACS Catal.*, **4**, 4498–4503.

68. Hamel, A., Sacco, M., Mnasri, N., et al. (2014) Micelles into glycerol solvent: overcoming side reactions of glycerol. *ACS Sustain. Chem. Eng.*, **2**, 1353–1358.

69. Hamid, M., Slatford, P.A. and Williams, J.M.J. (2007) Borrowing hydrogen in the activation of alcohols. *Adv. Synth. Catal.*, **349**, 1555–1575.

70. Dobereiner, G.E. and Crabtree, R.H. (2010) Dehydrogenation as a substrate-activating strategy in homogeneous transition-metal catalysis. *Chem. Rev.*, **110**, 681–703.

71. Azua, A., Mata, J.A. and Peris, E. (2011) Iridium NHC based catalysts for transfer hydrogenation processes using glycerol as solvent and hydrogen donor. *Organometallics*, **30**, 5532–5536.

72. Yadav, D.K.T., Rajak, S.S. and Bhanage, B.M. (2014) *N*-Arylation of indoles with aryl halides using copper/glycerol as a mild and highly efficient recyclable catalytic system. *Tetrahedron Lett.*, **55**, 931–935.

73. Ricordi, V.G., Freitas, C.S., Perin, G., et al. (2012) Glycerol as a recyclable solvent for copper-catalyzed cross-coupling reactions of diaryl diselenides with aryl boronic acids. *Green Chem.*, **14**, 1030–1034.

74. Mesquita, K.D., Waskow, B., Schumacher, R.F., et al. (2014) Glycerol/hypophosphorous acid and PhSeSePh: an efficient and selective system for reactions in the carbon–carbon double bond of (*E*)-chalcones. *J. Braz. Chem. Soc.*, **25**, 1261–1269.

75. Carmona, R.C., Schevciw, E.P., de Albuquerque, J.L.P., et al. (2013) Joint use of microwave and glycerol–zinc(II) acetate catalytic system in the synthesis of 2-pyridyl-2-oxazolines. *Green Process Synth.*, **2**, 35–42.

76. Bhojane, J.M., Sarode, S.A. and Nagarkar, J.M. (2016) Nickel–glycerol: an efficient, recyclable catalysis system for Suzuki cross coupling reactions using aryl diazonium salts. *New J. Chem.*, **40**, 1564–1570.

77. Chahdoura, F., Favier, I., Pradel, C., et al. (2015) Palladium nanoparticles stabilised by PTA derivatives in glycerol: synthesis and catalysis in a green wet phase. *Catal. Commun.*, **63**, 47–51.

78. González-Liste, P.J., Cadierno, V. and García-Garrido, S.E. (2015) Catalytic rearrangement of aldoximes to primary amides in environmentally friendly media under thermal and microwave heating: another application of the bis(allyl)-ruthenium(IV) dimer [{RuCl(μ-Cl)(η^3:η^3-C$_{10}$H$_{16}$)}$_2$]. *ACS Sustain. Chem. Eng.*, **3**, 3004–3011.

2

Aromatic Bio-Based Solvents

Egid B. Mubofu, James Mgaya, and Joan J. E. Munissi

Chemistry Department, University of Dar es Salaam, Dar es Salaam, Tanzania

2.1 Introduction

The development of science, technology and scientific research have for a long time depended on petroleum-based solvents. However, with the diminishing supply of petroleum due to overexploitation, high demand and the environmental effects associated with harnessing these non-renewable resources, there has come the demand that greener alternative solvents are made available in large supply. Bio-based solvents from easily accessible agricultural products and by-products are therefore an interesting venture, such solvents containing molecular scaffolds that offer diversity in terms of their applications, manipulations and prospects. In this chapter, we focus on alkylresorcinols (AR) and cashew nut shell liquid (CNSL) as bio-based solvents from agricultural produce and waste. The chapter highlights the possible biological sources for bio-based aromatics, such as cereals and cashew nut shell wastes, and the techniques for their extraction. The possible potential applications of these extracts as solvents or reagents in the production of functional or platform chemicals are also introduced. The competitive advantage of utilizing agricultural by-products or wastes, such as cereal bran and cashew nut shells, as renewable bio-resources for the production of aromatic bio-based solvents is manifested in non-interference with food supply and their contributions towards waste minimization. Thus, this chapter centres its discussion on

Bio-Based Solvents, First Edition. Edited by François Jérôme and Rafael Luque.
© 2017 John Wiley & Sons Ltd. Published 2017 by John Wiley & Sons Ltd.

aromatic bio-based solvents from agricultural by-products and waste materials, and we place emphasis on resorcinolic lipids and cashew nut shell liquid.

2.2   Resorcinolic Lipids

2.2.1   General Description

Resorcinolic lipids (also known as alkylresorcinols) are phenolic lipids that are produced primarily by plants, fungi and bacteria [1]. Chemically, they comprise 1,3-dihydroxy-5-alkylbenzene homologues with odd-numbered hydrocarbon side chains in the range of 15–25 carbon atoms [2]. This general chemical structure in which the non-isoprenoid side chain is attached to the phenolic ring makes resorcinolic lipids amphiphilic in nature. They are for simplicity considered as fatty acids, with the carboxyl group replaced by the phenolic ring. In most cases (especially in cereals), they occur as mixtures of at least several homologues, possessing different chain lengths and/or degrees of unsaturation (mono-unsaturated and/or di-unsaturated) [3]. The position of the double bonds in the alkyl side chain depends on its length. For instance, in C_{15} homologues, the double bonds occur at C8, C11 and C14 carbon atoms. For longer-chain homologues, the double bonds are localized at other carbon atoms, with each double bond separated from another by a methylene group. Oxygenated chain analogues (such as a keto group on the alkyl chain) have also been reported. However, in most cases, the alkyl chain is straight and the phenolic moieties remain unmodified (Figure 2.1) [1, 4–6].

2.2.2   Occurrence of Alkylresorcinols

Alkylresorcinols are the most abundant in nature when compared to other major classes of phenolic lipids, such as alkylphenols, anacardic acids and alkylcatechols. Over 100 various resorcinolic lipid homologues are known to occur in plant and microorganisms [1]. Kozubek and Tyman have summarized various plant families that have been identified as sources of alkylresorcinols [1]. These include the Ginkgoaceae, Anacardiaceae, Proteaceae, Myrsinaceae, Primulaceae, Myristicaceae, Iridaceae, Araceae, Compositae, Leguminoseae and Graminae families. It was also established that alkylresorcinols occur in algae (Chlorophycae, Sargassaceae), fungi (Basidiomycetes), bacteria (Hypnomycetes, Actinomycetales, Pseudomonales, Eubacteriales) and animals (Haliclonidae) [1].

Plants that accumulate high amounts of resorcinolic lipids include rye ($720–761\ \mu g\ g^{-1}$), triticale ($439–647\ \mu g\ g^{-1}$) and wheat ($489–1429\ \mu g\ g^{-1}$) [2], and low amounts have been found in barley (up to $54\ \mu g\ g^{-1}$) [7]. Additionally, there are reports of their occurrence in rice [8], rye seedlings [9], mango latex [10] and peel [11]. In cereals, alkylresorcinols are concentrated in the bran milling fraction and are generally not found in refined flour or in refined products from

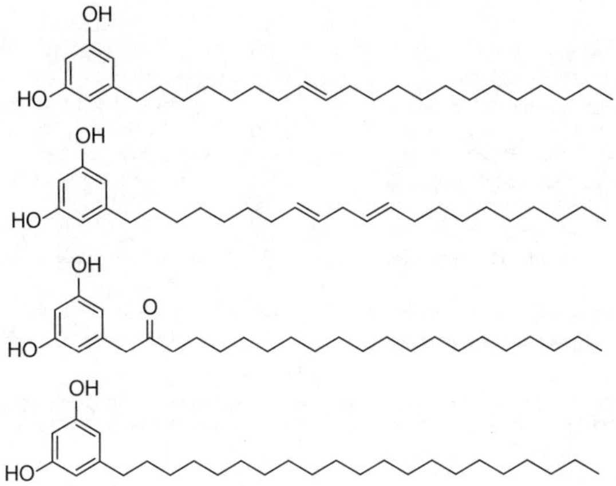

Figure 2.1 General structures of alkylresorcinols and oxygenated chain analogues.

cereals [12]. Figure 2.2 illustrates some of the common alkylresorcinols that occur in cereals [7, 13, 14].

2.2.3 Extraction of Alkylresorcinols

2.2.3.1 Solvent Extraction

Alkylresorcinols are miscible with hydrophobic solvents such as acetone, ethyl acetate, methanol, ethanol, diethyl ether, chloroform, cyclohexane and *n*-hexane. A simple static solvent extraction of alkylresorcinols is conducted at room temperature for 16–24 hours. The extraction time is shorter (2–6 hours) when Soxhlet apparatus is used for extraction [15]. Soxhlet extraction of alkylresorcinols using various hydrophobic solvents yields the highest amounts of total resorcinolic lipids especially in acetone and ethyl acetate extracts compared to other solvents [15]. The amount of solvent used during extraction may differ, but generally the ratio of sample to solvent is mostly between 0.02 and 0.03 (w/v) [14]. Extraction of alkyl-resorcinols from cereal grains can be done on whole grains with or without milling. Milling of grains decreases the extraction time; however, it increases the amount of co-extracted materials that later complicate the chromatographic analysis and lowers purity.

2.2.3.2 Accelerated Solvent Extraction

Accelerated solvent extraction (ASE) is an extraction technique that combines elevated temperatures and pressures with liquid solvents during the extraction process (Figure 2.3). Elevated pressures (>1000 psi) allow for solvents to be heated

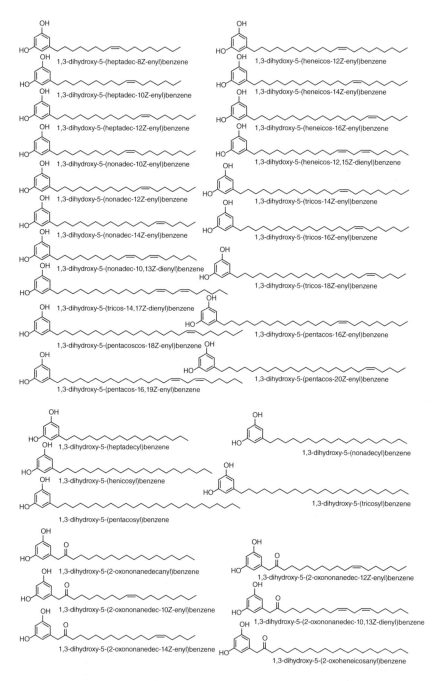

Figure 2.2   Structures of common alkylresorcinols found in cereals [7, 13, 14].

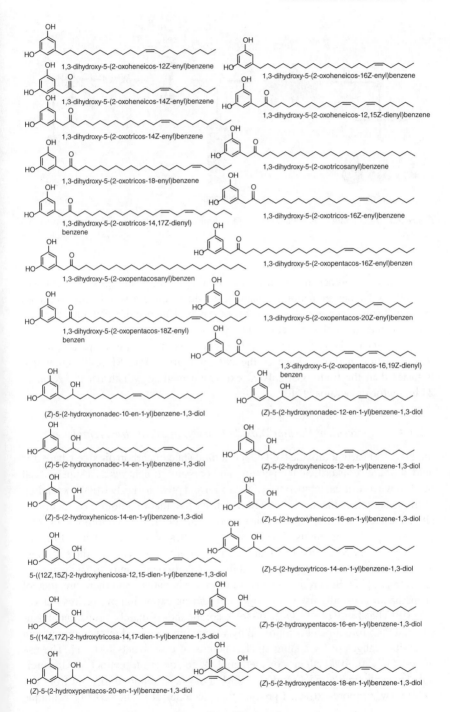

1,3-dihydroxy-5-(2-oxoheneicos-12Z-enyl)benzene

1,3-dihydroxy-5-(2-oxoheneicos-16Z-enyl)benzene

1,3-dihydroxy-5-(2-oxoheneicos-14Z-enyl)benzene

1,3-dihydroxy-5-(2-oxoheneicos-12,15Z-dienyl)benzene

1,3-dihydroxy-5-(2-oxotricos-14Z-enyl)benzene

1,3-dihydroxy-5-(2-oxotricosanyl)benzene

1,3-dihydroxy-5-(2-oxotricos-18-enyl)benzene

1,3-dihydroxy-5-(2-oxotricos-16Z-enyl)benzene

1,3-dihydroxy-5-(2-oxotricos-14,17Z-dienyl) benzene

1,3-dihydroxy-5-(2-oxopentacos-16Z-enyl)benzen

1,3-dihydroxy-5-(2-oxopentacosanyl)benzen

1,3-dihydroxy-5-(2-oxopentacos-20Z-enyl)benzen

1,3-dihydroxy-5-(2-oxopentacos-18Z-enyl) benzen

1,3-dihydroxy-5-(2-oxopentacos-16,19Z-dienyl) benzen

(Z)-5-(2-hydroxynonadec-10-en-1-yl)benzene-1,3-diol

(Z)-5-(2-hydroxynonadec-12-en-1-yl)benzene-1,3-diol

(Z)-5-(2-hydroxynonadec-14-en-1-yl)benzene-1,3-diol

(Z)-5-(2-hydroxyhenicos-12-en-1-yl)benzene-1,3-diol

(Z)-5-(2-hydroxyhenicos-14-en-1-yl)benzene-1,3-diol

(Z)-5-(2-hydroxyhenicos-16-en-1-yl)benzene-1,3-diol

5-((12Z,15Z)-2-hydroxyhenicosa-12,15-dien-1-yl)benzene-1,3-diol

(Z)-5-(2-hydroxytricos-14-en-1-yl)benzene-1,3-diol

5-((14Z,17Z)-2-hydroxytricosa-14,17-dien-1-yl)benzene-1,3-diol

(Z)-5-(2-hydroxypentacos-16-en-1-yl)benzene-1,3-diol

(Z)-5-(2-hydroxypentacos-20-en-1-yl)benzene-1,3-diol

(Z)-5-(2-hydroxypentacos-18-en-1-yl)benzene-1,3-diol

Figure 2.2   (*Continued*)

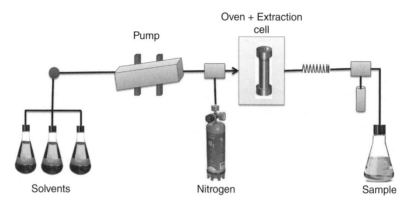

Figure 2.3   Accelerated solvent extraction. (A colour version of this figure appears in the plate section.)

at temperatures higher than their normal boiling point, resulting in fast, efficient extractions. Higher temperatures ensure faster diffusion rate, which improves mass transfer. It also increases the capacity of solvents to solubilize the analytes. A study carried out on the extraction of alkylresorcinols in food products containing uncooked and cooked wheat using ASE revealed results that are comparable to current extraction methods. However, the extraction time with ASE is 40 min, which is faster than the traditional solvent extraction method, which usually requires 24 hours for raw grains [16].

2.2.3.3   *Supercritical Carbon Dioxide Extraction of Alkylresorcinols*

Supercritical carbon dioxide (sc-CO_2) is a fluid state of carbon dioxide where it is held at or above its critical temperature and pressure. It behaves as a supercritical fluid above its critical temperature (304.25 K) and pressure (7.39 MPa), expanding to fill its container like a gas but with a density like that of a liquid. Extraction of alkylresorcinols with sc-CO_2 is considered as an alternative to traditional organic solvent extraction methods. The gas-like properties of sc-CO_2, such as low surface tension and viscosity, facilitate penetration of the solvent into the substrate, whereas the liquid-like properties solubilize the compounds and extract them from the biomass [17]. Supercritical CO_2 is non-polar in nature, and therefore ethanol or methanol is usually used as a co-solvent during extraction of alkylresorcinols [18, 19]. The ability of sc-CO_2 to dissolve compounds is mainly a function of density and therefore it can be improved by varying pressure and temperature within accessible ranges or by adding small amounts of co-solvents [20, 21]. The use of sc-CO_2 with 10% ethanol or methanol as co-solvent to extract alkylresorcinols from rye bran at 55°C and 15–30 MPa has been reported to yield from 8% to 80% (w/w) more extracted product than acetone [18]. The sc-CO_2 technique

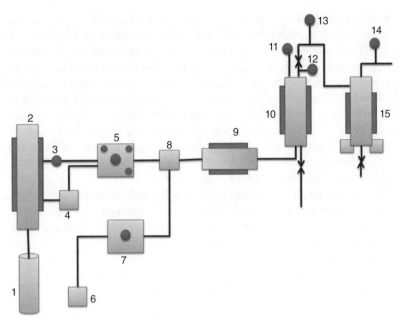

Figure 2.4 Schematic diagram of the supercritical extraction system: 1, CO_2 cylinder; 2, cooling heat exchanger; 3, flowmeter; 4, cooling bath; 5, CO_2 pump; 6, co-solvent reservoir; 7, co-solvent pump; 8, mixer; 9, heat exchanger; 10, extraction vessel; 11, pressure gauge; 12, temperature sensor; 13, automated back-pressure pump; 14, pressure gauge; 15, cyclone separator [22]. (A colour version of this figure appears in the plate section.)

has also been applied in the extraction and fractionation of alkylresorcinols from triticale bran. In such extraction, saturated and unsaturated alkylresorcinol homologues of C15:0, C17:0, C19:0, C19:1, C21:0, C21:1, C23:0 and C25:0 have been detected in all samples, with over 98% of the triticale alkylresorcinols and over 95% of the wheat alkylresorcinols being extracted under the sc-CO_2 conditions used [22]. In comparison to the traditional solvent extraction methods, sc-CO_2 extraction (Figure 2.4) is an attractive technique because the solvent is non-toxic, green and non-flammable and the method has a faster extraction time [23, 24].

2.2.4   Scientific Interest in Alkylresorcinols

Resorcinolic lipids are reported to have a wide range of scientific interest in various fields of study and application, such as pharmacology, biomedicine and biotechnology. The amphiphilic nature of alkylresorcinols is an important aspect with regard to their analysis, absorption, metabolism and potential bioactivities. They have been reported to possess antimicrobial, antiparasitic and cytotoxic activity. The biological activities of alkylresorcinols have recently been intensively reviewed [25]. The compounds have been reported to affect oxidation processes, such as preventing Fe^{2+}-induced peroxidation of fatty acids and phospholipids in liposomal membranes as well as in the autoxidation processes in triglycerides and fatty acids [26, 27]. They have also been reported to be effective in protecting the erythrocyte membrane against hydrogen-peroxide-induced oxidation [28]. Furthermore, alkylresorcinols can interact with DNA in the body's cells, resulting in the absence of mutagenic, carcinogenic and co-carcinogenic effects [29], which is a desirable property in pharmacology and medicine. The ability of alkylresorcinols to interact with cell membranes and DNA structures, as well as their cytotoxic activities, makes them potential agents for inhibition of bacterial, fungal, protozoan and parasite growth.

Once absorbed in the body, alkylresorcinols can be detected in the human plasma and as metabolites in urine. Through understanding the kinetics of their absorption, metabolism and storage in the body, these compounds can serve as potential biomarkers for intake of wholegrain wheat and rye [30–32]. Studies on alkylresorcinols in human plasma and urinary metabolites strongly suggest that these compounds are responsive to wholegrain wheat and rye intake and are correlated with various measures of alkylresorcinol consumption [33].

2.2.4.1   Fine Chemicals from Alkylresorcinols

Alkylresorcinols have found wide applications in various fields, including pharmaceuticals, agrochemicals, dyes, cosmetics, food additives, novolacs in the electronics industry, high-value photoactive resins, additives in the plastics industry, UV blockers, and antioxidants and stabilizers in the rubber industry [34]. Alkylresorcinols have also been used in the preparation of modified phenol–formaldehyde resins, making it possible to manufacture plywood with increased and improved properties, especially bond quality [35]. As a result of their widespread occurrence in nature and their chemical structure, which is intrinsically easy to manipulate, alkylresorcinols with various chain lengths can be obtained from the cereal bran materials produced as agricultural by-products and available in bulk. Consequently, valorization of the bran to produce several bio-based aromatics, such as orsellinic acid, 2,6-dihydroxy-4-methylbenzoic acid, 5-methylresorcinol (orcinol), 2,6-dimethoxy-4-methylphenol, 3,5-dimethoxytoluene, 3-hydroxy-5-methoxytoluene, 2,5-dimethylresorcinol and 4,5-dimethylresorcinol to mention just a few (Figure 2.5), on a large scale would be advantageous over embarking on individual separation of these chemical entities from their original natural sources scattered over many different species of plants, lichens, fungi and algae [36–38].

Alkylresorcinols have been utilized as precursors in the synthesis of various molecules, such as the non-natural cannabinoids, grifolin, neogrifolin, schizopeltic acid, cellectochlorine, resveratrol, coumarins, isoflavanoids, etc.

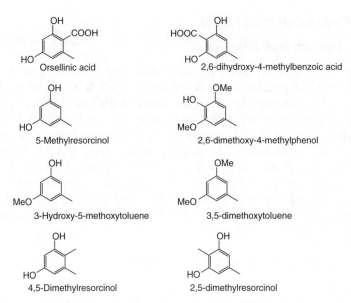

Figure 2.5   Some examples of bio-based aromatics that can be derivatized from the alkylresorcinols occurring in cereals.

[39–41]. 5-Methylresorcinol has also been utilized efficiently as a starting material in the total synthesis of the natural product, daurichromenic acid (Scheme 2.1), a highly potent anti-HIV agent [42].

5-Methylresorcinol, which is also a naturally occurring alkylresorcinol, can be envisaged to be produced by partial degradation of the long alkyl tail of the alkylresorcinol. This aromatic molecule has recently become a chemical of choice to replace resorcinol in many reactions. It was established that the presence of the methyl group at position 5 in the aromatic ring contributes to a significant reduction in toxicity by almost four times compared to resorcinol [43]. Additionally, 5-methylresorcinol is comparably less expensive. Owing to the aforementioned properties, 5-methylresorcinol was recently used as a precursor for the preparation of carbon aerogels (replacing resorcinol), which are porous carbon materials characterized by low density and high surface area [43, 44]. Carbon aerogels are applicable as catalyst support, materials for electrodes and adsorbents.

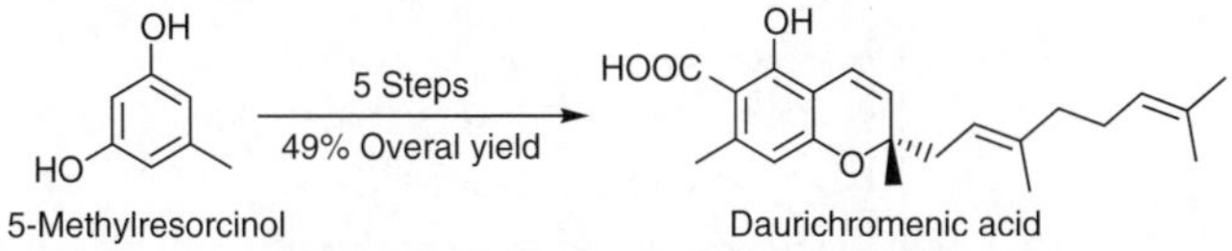

Scheme 2.1   Synthesis of daurichromenic acid using an alkylresorcinol as the starting material [42].

2.3 Cashew Nut Shell Liquid

2.3.1 Description and Occurrence

Cashew nut shell liquid (CNSL) is a dark reddish brown viscous liquid, found in a soft honeycomb structure inside the cashew nut shell. It is a by-product of the cashew industry obtained during the process of removing the cashew kernel from the nut. Cashew nuts consist of 35–45% kernel and around 55–65% of shells. The shells contain 15–30% CNSL. Worldwide, the annual production of raw cashew nuts is about 3.6 million tons, the top five producers being Vietnam (1.1 million tons), Nigeria (0.95 million tons), India (0.75 million tons), Côte d'Ivoire (0.45 million tons) and Brazil (0.11 million tons) [45]. Thus, CNSL represents the largest readily available bio-resource of naturally occurring long-chain alkenylphenolic compounds, the major components of CNSL being anacardic acid, cardanol, cardol and 2-methylcardol (Figure 2.6).

2.3.2 Extraction of Cashew Nut Shell Liquid

Basically, there are three methods used to extract CNSL from the shells: solvent extraction, thermal extraction and mechanical extraction. Depending on the mode of extraction, CNSL is classified into technical and natural CNSL. Natural CNSL is obtained by cold solvent extraction and contains anacardic acid (60–65%), cardanol (10–15%), cardol (20%) and traces of 2-methylcardol [45]. These compositions may vary depending on the geographical locations at which the cashew plant is grown [46, 47]. For technical CNSL, the extraction involves higher temperatures and thus most of the anacardic acid is decarboxylated during roasting of the shells. Technical CNSL contains cardanol (60–65%), cardol (15–20%), polymeric materials (10%) and traces of 2-methylcardol [45].

2.3.2.1 Solvent Extraction of CNSL

Solvent extraction is the best approach when CNSL rich in anacardic acid is needed. The solvent extraction techniques that are frequently employed for the

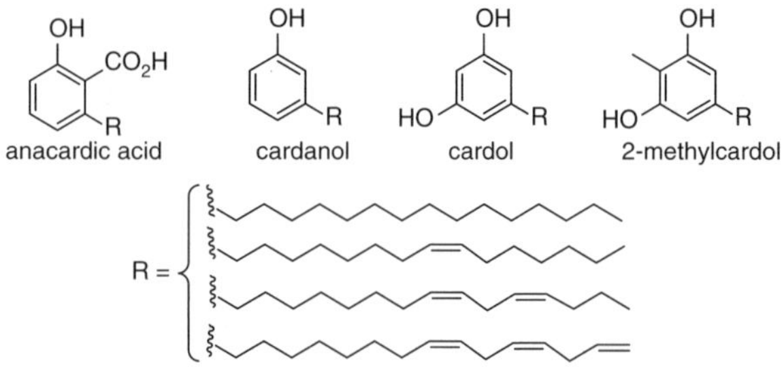

Figure 2.6 Components of cashew nut shell liquid.

extraction of CNSL include static solvent extraction [48], Soxhlet extraction [49, 50], ultrasonic extraction [48], two-step extraction [51], subcritical water (SCW) extraction [51] and supercritical carbon dioxide [52, 53]. The various solvents used in the extraction of CNSL include petroleum ether, carbon tetrachloride, hexane, cyclohexane, diethyl ether, xylene, ethyl acetate, toluene, ethanol, methanol and acetone. Polar solvents have been reported to give higher amounts of extracted CNSL than non-polar solvents [54].

2.3.2.2 Thermal Extraction of CNSL

In thermal extraction, CNSL is obtained by roasting shells, using methods that include open pan roasting, drum roasting and hot oil roasting. Extraction by roasting is the traditional method of removing CNSL. The application of heat to the nut releases CNSL and makes the shell brittle to facilitate the extraction of the kernel by breaking the shell. The extraction is performed at higher temperatures (above 180°C) at which decarboxylation of anacardic acid occurs to form cardanol. The thermal extraction approach is therefore suitable when cardanol-rich CNSL is desired. Among the three methods of thermal extraction mentioned above, the hot oil method is more suitable considering the viability of CNSL collection [55]. The principle employed in the hot oil method is that the shells, when immersed in the tank containing oil at high temperature, will lose their CNSL and hence increase the volume of the oil in the tank. The simple equipment for the hot oil method consists of a tank of CNSL heated to a temperature of 185–190°C by a furnace underneath, and a wire basket used to hold the nuts for immersion into the tank. Extraction by the hot oil method produces around 6–12% CNSL by weight of a nut [55].

2.3.2.3 Mechanical Extraction of CNSL

In the mechanical method (screw press method), the raw cashew nut shells are put into a hydraulic screw press and then a high pressure is exerted in order to release CNSL from the shells. Extraction of CNSL by pressing gives 20–21% of CNSL [56].

2.3.3 Scientific Interest in Cashew Nut Shell Liquid

CNSL is rich in alkenylphenols and anacardic acid, which have attracted much interest by many scientists. This interest is substantiated by the fact that CNSL does not compete with food production and is produced from solid waste. In addition, CNSL has found application in several sectors, including the polymer industry, for biomedical use and in the synthesis of fine industrial chemicals. Some researchers have used natural CNSL as a templating agent to synthesize mesoporous materials for supporting metal catalysts and enzymes. Anacardic acids have been reported to be green capping agents for the synthesis of nanoparticles. The applications of these aromatic structures and their manipulation to give other important products are discussed in detail in the subsequent subsections.

2.3.3.1   Biological Activity of CNSL

Anacardic acid (see Figure 2.6), the major component of cold solvent-extracted CNSL, possesses antibacterial, antitumour and antioxidant activities and has served as a synthon for the preparation of a variety of biologically active compounds. As an antibacterial agent, the activity of anacardic acid, possessing an alkyl triene side chain, against *Streptococcus mutans* (ATCC 25175) and *Staphylococcus aureus* (ATCC 12598) was 2048 and 64 times more effective than salicylic acid [57, 58]. As an antitumour agent, anacardic acid has been reported to affect multiple steps of tumour angiogenesis, thereby bringing about tumour growth inhibition [59]. It has been found that anacardic acid significantly suppresses vascular endothelial growth factor (VEGF) induced cell proliferation, migration and adhesion and capillary-like structure formation of primary cultured human umbilical vascular endothelial cells (HUVECs) without detectable toxicity [59]. The antioxidant property of anacardic acid is exerted by preventing the generation of superoxide radicals through inhibition of the xanthine oxidase (EC 1.1.3.22, Grade IV) enzyme. It inhibits various prooxidant enzymes involved in the production of the reactive oxygen species by chelating divalent metal ions such as Fe^{2+} or Cu^{2+} [60]. The anticancer activity and many other medicinal properties of anacardic acid have been reviewed elsewhere [57, 61].

2.3.3.2   CNSL in the Polymer Industry

CNSL-based polymers and resins have numerous industrial applications, such as friction linings, paints and varnishes, laminating resins, rubber compounding resins, cashew cements, polyurethane-based polymers, surfactants, epoxy resins, surface coatings, adhesives, flame retardants and anticorrosive paint [62–64]. Cardanol polysulfide is used as a vulcanizing agent for natural rubber [65]. Cardanol-based novolac-type phenolic resins are blended with commercial epoxy or isocyanate monomers to form thermoset polymers [66]. Monomers from anacardic acid (anacardanyl acrylate and anacardanyl methacrylate) were synthesized and utilized to prepare molecularly imprinted polymers [67]. The structural framework of the alkenylphenols found in CNSL is well suited as a source of vinylic monomers for the preparation of molecularly imprinted polymers for the development of various bioanalytical technologies.

2.3.3.3 Fine Chemicals from CNSL

The structure of the components of CNSL presents a multitude of possibilities for transformation into other more useful industrial chemicals. The presence of hydroxyl groups and double bonds makes cardanol a suitable substrate for the synthesis of bifunctional monomers. This has been demonstrated by the recent synthesis of kairomone components via isomerization and metathesis reactions (Scheme 2.2) using cardanol and anacardic acid as precursors [68–70].

Ethenolysis reactions of cardanol, anacardic acid and cardol give 1-octene, which is used as an industrial intermediate, primarily in the production of polyethylene resins. 1-Octene is also used for the synthesis of linear and branched aldehydes by hydroformylation. The ethenolysis reaction of cardanol produces (*E*)-3-(non-8-enyl)phenol (Scheme 2.3), the hydrogenation of which produces 3-nonylphenol. The latter is thought to be a good replacement for 4-nonylphenol, which has been banned in the European Community because of its endocrine-disrupting properties [68, 71]. The butenolysis reaction of cardanol produces (*E*)-3-(non-8-en-1-yl)phenol, which is a precursor for the synthesis of a detergent, sodium (*E*)-2-hydroxy-6-(non-8-en-1-yl)benzenesulfonate (Scheme 2.4) [68].

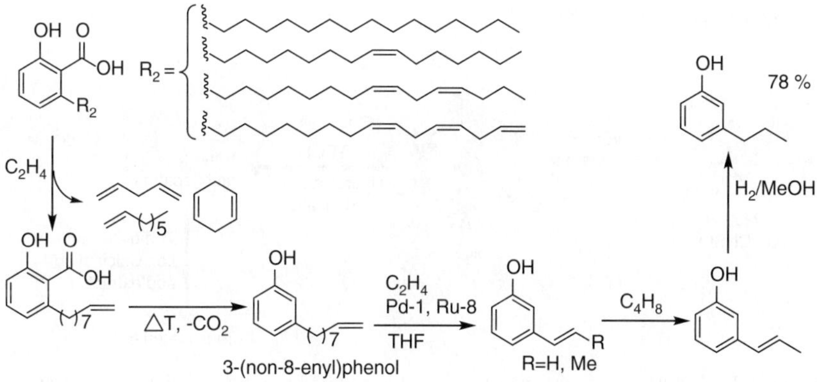

Scheme 2.2 Synthesis of 3-propylphenol via isomerizing metathesis of 3-(non-8-enyl)phenol [69].

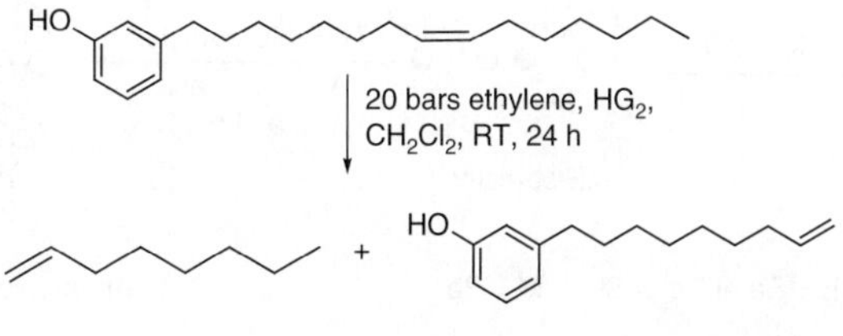

Scheme 2.3 Synthesis of 1-octene and (*E*)-3-(non-8-enyl)phenol [69].

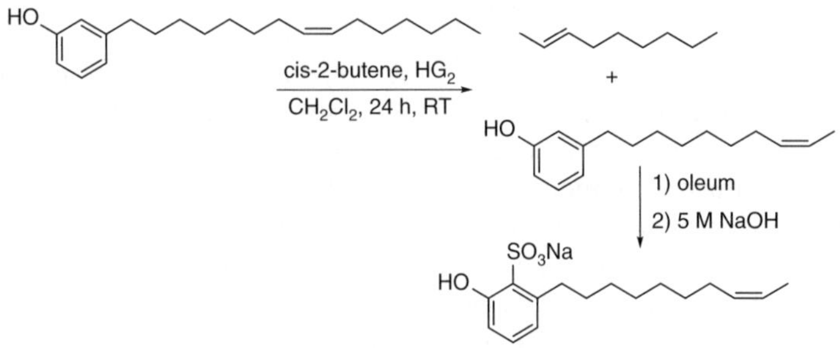

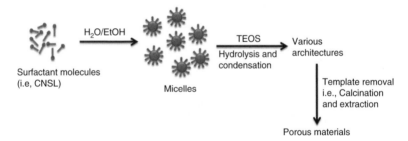

Scheme 2.4   Synthesis of a detergent from cardanol [68].

2.3.3.4   Other Uses of CNSL

CNSL has been reported elsewhere to be a green template for the preparation of mesoporous materials (Figure 2.7). The prepared materials have pores with diameter of up to 25 nm suitable for supporting metal catalysts and enzymes for heterogeneous catalysis [72, 73]. Furthermore, anacardic acid has been reported as a green capping agent for the synthesis of anacardic acid-capped PbS and PbSe nanoparticles (Figure 2.8) [74].

Figure 2.7   Preparation of porous materials employing CNSL as a templating agent. (A colour version of this figure appears in the plate section.)

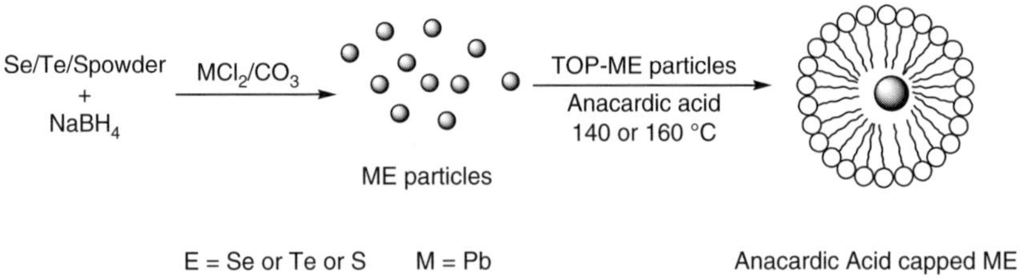

Figure 2.8   Preparation of anacardic acid-capped PbS and PbSe nanoparticles.

2.4 Conclusion

Aromatic bio-based solvents represent an important platform not only for the development of new technologies and advances in science, but also as an opportunity for profitable investment in bio-based chemical industries. The demand for and consumption of aromatic bio-based solvents will continue to rise in the coming years because they meet the current demands for environmental protection, are inexpensive and are green alternatives to the existing toxic solvents, as exemplified in some of the products discussed herein, such as 5-methylresorcinol and 3-nonylphenol, which have replaced their toxic counterparts, resorcinol and 4-nonylphenol, respectively. Owing to their intrinsic biological properties, aromatic bio-based solvents are at the competitive cutting edge as the most sought-after building blocks for the construction of chemical libraries with varied applications. The attractiveness of opting for bio-based solvents lies in their widespread and substantial occurrence in agricultural materials that will not jeopardize food availability and accessibility. This means that they are important raw materials for the production of aromatic bio-based solvents for applications such as aromatic building blocks, active ingredients or advanced intermediates in chemical processes.

References

1. Kozubek, A. and Tyman, J.H.P. (1999) Resorcinolic lipids, the natural non-isoprenoid phenolic amphiphiles and their biological activity. *Chem. Rev.*, **99**, 1–24.
2. Ross, A.B., Shepherd, M.J., Schüpphaus, M.S., *et al.* (2003) Alkylresorcinols in cereals and cereal products. *J. Agric. Food Chem.*, **51**, 4111–4118.
3. Kozubek, A. and Tyman, J.H.P. (1995) Cereal grain resorcinolic lipids: mono- and dienoic homologues are present in rye grains. *Chem. Phys. Lipids*, **78**, 29–35.
4. Suzuki, Y., Esumi, Y., Saito, T., *et al.* (1998) Identification of 5-n-(2'-oxo)alkylresorcinols from etiolated rice seedlings. *Phytochemistry*, **47**, 1247–1252.
5. Suzuki, Y., Esumi, Y. and Yamaguchi, I. (1999) Structures of 5-alkylresorcinol-related analogues in rye. *Phytochemistry*, **52**, 281–289.
6. Seitz, L.M. (1992) Identification of 5-(2-oxoalkyl)resorcinols and 5-(2-oxoalkenyl) resorcinols in wheat and rye grains. *J. Agric. Food Chem.*, **40**, 1541–1546.
7. Zarnowski, R., Suzuki, Y., Yamaguchi, I. and Pietr, S.J. (2002) Alkylresorcinols in barley (*Hordeum vulgare* L. *distichon*) grains. *Z. Naturforsch. C*, **57**, 57–62.
8. Suzuki, Y., Esumi, Y., Hyakutake, H., *et al.* (1996) Isolation of 5-(8'Z-heptadecenyl) resorcinol from etiolated rice seedlings as an antifungal agent. *Phytochemistry*, **41**, 1485–1489.
9. Deszcz, L. and Kozubek, A. (2000) Higher cardol homologs (5-alkylresorcinols) in rye seedlings. *Biophys. Acta*, **1483**, 241–250.
10. Bandyopadhyay, C., Gholap, A.S. and Mamdapur, V.R. (1985) Characterization of alkenyl-resorcinol in mango (*Mangifera indica* L.) latex. *J. Agric. Food Chem.*, **33**, 377–379.
11. Knödler, M., Berardini, N., Kammerer, D.R., *et al.* (2007) Characterization of major and minor alk(en)ylresorcinols from mango (*Mangifera indica* L.) peels by high-performance

liquid chromatography/atmospheric pressure chemical ionization mass spectrometry. *Rapid Commun. Mass Spectrom.*, **21**, 945–951.

12. Al-Ruqaie, I. and Lorenz, K. (1992) Alkylresorcinols in extruded cereal bran. *Cereal Chem.*, **69**, 472–475.

13. Landberg, R., Kamal-Eldin, A., Andersson, R. and Åman, P. (2006) Alkylresorcinol content and homologue composition in durum wheat (*Triticum durum*), kernels and pasta products. *J. Agric. Food Chem.*, **54**, 3012–3014.

14. Ross, A.B., Åman, P., Andersson, R. and Kamal-Eldin, A. (2004) Chromatographic analysis of alkylresorcinols and their metabolites. *J. Chromatogr. A*, **1054**, 157–164.

15. Zarnowski, R. and Suzuki, Y. (2004) Expedient Soxhlet extraction of resorcinolic lipids from wheat grains. *J. Food Compos. Anal.*, **17**, 649–663.

16. Holt, M.D., Moreau, R.A., DerMarderosian, A., *et al.* (2012) Accelerated solvent extraction of alkylresorcinols in food products containing uncooked and cooked wheat. *J. Agric. Food Chem.*, **60**, 4799–4802.

17. McKenzie, L.C., Thompson, J.E., Sullivan, R. and Hutchison, J.E. (2004) Green chemical processing in the teaching laboratory: a convenient liquid CO_2 extraction of natural products. *Green Chem.*, **6**, 355–358.

18. Francisco, J.D.C., Danielsson, B., Kozubek, A. and Dey, E.S. (2005) Extraction of rye bran by supercritical carbon dioxide: influence of temperature, CO_2, and co-solvent flow rates. *J. Agric. Food Chem.*, **53**, 7432–7437.

19. Sairam, P., Ghosh, S., Jena, S., *et al.* (2012) Supercritical fluid extraction (SFE) – an overview. *Asian J. Res. Pharm. Sci.*, **2**, 112–120.

20. Zaidul, I.S.M., Norulaini, N.A.N., Omar, A.K. and Smith, Jr.,, R.L. (2006) Supercritical carbon dioxide (SC-CO_2) extraction and fractionation of palm kernel oil from palm kernel as cocoa butter replacers blend. *J. Food Eng.*, **73**, 210–216.

21. Rebolleda, S., Beltrán, S., Sanz, M.T., *et al.* (2013) Extraction of alkylresorcinols from wheat bran with supercritical CO_2. *J. Food Eng.*, **119**, 814–821.

22. Athukorala, Y., Hosseinian, F.S. and Mazza, G. (2010) Extraction and fractionation of alkyl-resorcinols from triticale bran by two-step supercritical carbon dioxide. *LWT – Food Sci. Technol.*, **43**, 660–665.

23. Shilpi, A., Shivhare, U.S. and Basu, S. (2013) Supercritical CO_2 extraction of compounds with antioxidant activity from fruits and vegetables waste – a review, focusing on modern food industry. *Focus. Mod. Food Ind.*, **2**, 43–62.

24. Gunenc, A., HadiNezhad, M., Farah, I., *et al.* (2015) Impact of supercritical CO_2 and tra-ditional solvent extraction systems on the extractability of alkylresorcinols, phenolic profile and their antioxidant activity in wheat bran. *J. Funct. Foods*, **12**, 109–119.

25. Stasiuk, M. and Kozubek, A. (2010) Biological activity of phenolic lipids. *Cell. Mol. Life Sci.*, **67**, 841–860.

26. Korycinska, M., Czelna, K., Jaromin, A. and Kozubek, A. (2009) Antioxidant activity of rye bran alkylresorcinols and extracts from whole-grain cereal products. *Food Chem.*, **116**, 1013–1018.

27. Kamal-Eldin, A., Pouru, A., Eliasson, C. and Åman, P. (2000) Alkylresorcinols as antiox-idants: hydrogen donation and peroxyl radical-scavenging effects. *J. Sci. Food Agric.*, **81**, 353–356.

28. Parikka, K., Rowland, I.R., Welch, R.W. and Wahala, K. (2006) In vitro antioxidant activity and antigenotoxicity of 5-n-alkylresorcinols. *J. Agric. Food Chem.*, **54**, 1646–1650.

29. Gasiorowski, K., Szyba, K., Brokos, B. and Kozubek, A. (1996) Antimutagenic activity of alkylresorcinols from cereal grains. *Cancer Lett.*, **106**, 109–115.

30. Landberg, R., Kamal-Eldin, A., Andersson, A., *et al.* (2008) Alkylresorcinols as biomarkers of whole-grain wheat and rye intake: plasma concentration and intake estimated from dietary records. *Am. J. Clin. Nutr.*, **87**, 832–838.

31. Van Dam, R.M. and Hu, F.B. (2008) Are alkylresorcinols accurate biomarkers for whole grain intake? *Am. J. Clin. Nutr.*, **87**, 797–798.

32. Landberg, R., Marklund, M., Kamal-Eldin, A. and Åman, P. (2014) An update on alkylresorcinols – occurrence, bioavailability, bioactivity and utility as biomarkers. *J. Funct. Foods*, **7**, 77–89.

33. Ross, A.B. (2012) Present status and perspectives on the use of alkylresorcinols as biomarkers of wholegrain wheat and rye intake. *J. Nutr. Metabol.*, **2012**, 1–12.

34. Pérez-Caballero, F., Peikolainen, A.-L., Koel, M., *et al.* (2008) Preparation of the catalyst support from the oil-shale processing by-product. *Open Petrol. Eng. J.*, **1**, 42–46.

35. Dziurka, D., Łęcka, J. and Mirski, R. (2009) The effect of modification of phenolic resin with alkylresorcinols and H_2O_2 on properties of plywood. *Acta Sci. Polon., Silv. Colend. Ratio Ind. Lign.*, **8**, 67–74.

36. Munissi, J.J.E. (2011) New polycyclic ketides and other metabolites from cultures of some Tanzanian marine fungi. Ph.D. Thesis, University of Dar es Salaam, Dar es Salaam, Tanzania.

37. Nolan, T.J., Keane, J. and Davidson, V.E. (1940) Chemical constituents of the lichen *Parmelia latissima* Fee. *Sci. Proc. R. Dublin Soc., Ser. A*, **22**, 237–239.

38. Baldermann, S., Yang, Z., Sakai, M., *et al.* (2009) Volatile constituents in the scent of roses. *Floricult. Ornam. Biotechnol.*, **3**, 89–97.

39. Huffman, J.W., Duncan, Jr., S.G., Wiley, J.L. and Martin, B.R. (1997) Synthesis and pharmacology of the 1′,2′-dimethylheptyl-Δ8-THC isomers: exceptionally potent cannabinoids. *Bioorg. Med. Chem. Lett.*, **7**, 2799–2804.

40. Ima-ye, K. and Kakisawa, H. (1973) Synthesis of grifolin and dihydrodeoxytauranin. *J. Chem. Soc., Perkin Trans. 1*, **1973**, 2591–2595.

41. Koch, K., Podlech, J., Pfeiffer, E. and Metzler, M. (2005) Total synthesis of alternariol. *J. Org. Chem.*, **70**, 3275–3276.

42. Kang, Y., Mei, Y., Du, Y. and Jin, Z. (2003) Total synthesis of the highly potent anti-HIV natural product daurichromenic acid along with its two chromane derivatives, rhododaurichromanic acids A and B. *Org. Lett.*, **5**, 4481–4484.

43. Pérez-Caballero, F., Peikolainen, A.-L., Uibu, M., *et al.* (2008) Preparation of carbon aerogels from 5-methylresorcinol–formaldehyde gels. *Micropor. Mesopor. Mater.*, **108**, 230–236.

44. Kreek, K., Kulp, M., Uibu, M., *et al.* (2014) Preparation of metal-doped carbon aerogels from oil shale processing by-products. *Oil Shale*, **31**, 185–194.

45. Food and Agriculture Organization of the United Nations (2013) *FAOSTAT. Food and agriculture data.* http://faostat3.fao.org (accessed June 2015).

46. Kumar, P.P., Paramashivappa, R., Vithayathil, P.J., *et al.* (2002) Process for isolation of cardanol from technical cashew (*Anacardium occidentale* L.) nut shell liquid. *J. Agric. Food Chem.*, **50**, 4705–4708.

47. Shoba, S.V. and Ravindranath, B. (1991) Supercritical carbon dioxide and solvent extraction of the phenolic lipids of cashew nut (*Anacardium occidentale*) shells. *J. Agric. Food Chem.*, **39**, 2214–2217.

48. Tyman, J.H.P., Johnson, R.A., Muir, M. and Rokhgar, R. (1989) The extraction of natural cashew nut shell liquid from the cashew nut (*Anacardium occidentale*). *J. Am. Oil Chem. Soc.*, **66**, 553–557.
49. Kumar, P.S., Kumar, N.A., Sivakumar, R. and Kaushik, C. (2009) Experimentation on solvent extraction of polyphenols from natural waste. *J. Mater. Sci.*, **44**, 5894–5899.
50. Idah, P.A., Simeon, M.I. and Mohammed, M.A. (2014) Extraction and characterization of cashew nut (*Anacardium occidentale*) oil and cashew shell liquid oil. *Acad. Res. Int.*, **5**, 50–54.
51. Yuliana, M., Thi, N.Y.T. and Ju, Y.H. (2012) Effect of extraction methods on characteristic and composition of Indonesian cashew nut shell liquid. *Ind. Crops Prod.*, **35**, 230–236.
52. Patel, R.N., Bandyopadhyay, S. and Ganesh, A. (2006) Extraction of cashew (*Anacardium occidentale*) nut shell liquid using supercritical carbon dioxide. *Bioresour. Technol.*, **97**, 847–853.
53. Smith, Jr., R.L., Malaluan, R.M., Setianto, W.B., *et al.* (2003) Separation of cashew (*Anacardium occidentale* L.) nut shell liquid with supercritical carbon dioxide. *Bioresour. Technol.*, **88** (1), 1–7.
54. Gandhi, T.S., Dholakiya, B.Z. and Patel, M.R. (2013) Extraction protocol for isolation of CNSL by using protic and aprotic solvents from cashew nut and study of their physico-chemical parameter. *Pol. J. Chem. Technol.*, **15** (4), 24–27.
55. Garkal, D.J. (2014) Review on extraction and isolation of cashew nut shell liquid. *Int. J. Innov. Eng. Res. Technol.*, **1** (1), 1–8.
56. Subbarao, C.N.V., Krishna Prasad, K.M.M. and Prasad, V.S.R.K. (2011) Review on applications, extraction, isolation and analysis of cashew nut shell liquid (CNSL). *Pharma Res. J.*, **6** (1), 21–41.
57. Hamad, F.B. and Mubofu, E.B. (2015) Potential biological applications of bio-based anacardic acids and their derivatives. *Int. J. Mol. Sci.*, **16**, 8569–8590.
58. Himejima, M. and Kubo, I. (1991) Antibacterial agents from the cashew *Anacardium occidentale* (Anacardiaceae) nut shell oil. *J. Agric. Food Chem.*, **39**, 418–421.
59. Wu, Y., He, L., Zhang, L., *et al.* (2011) Anacardic acid (6-pentadecylsalicylic acid) inhibits tumor angiogenesis by targeting Src/FAK/Rho GTPases signaling pathway. *J. Pharmacol. Exp. Ther.*, **339**, 403–411.
60. Kubo, I., Masuoka, N., Ha, T.J. and Tsujimoto, K. (2006) Antioxidant activity of anacardic acids. *Food Chem.*, **99**, 555–556.
61. Hemshekhar, M., Santhosh, M.S., Kemparaju, K. and Girish, K.S. (2011) Emerging roles of anacardic acid and its derivatives: a pharmacological overview. *Basic Clin. Pharmacol. Toxicol.*, **110**, 122–132.
62. Lubi, M.C. and Thachil, E.T. (2000) Cashew nut shell liquid (CNSL) – a versatile monomer for polymer synthesis. *Des. Monomers Polym.*, **3**, 123–153.
63. Voirin, C., Caillol, S., Sadavarte, N.V., *et al.* (2014) Functionalization of cardanol: towards biobased polymers and additives. *Polym. Chem.*, **5**, 3142–3162.
64. Balgude, D. and Sabnis, A.S. (2014) CNSL: an environment friendly alternative for the modern coating industry. *J. Coat. Technol. Res.*, **11** (2), 169–183.
65. Leerawan, K., Sawasdipuksa, N., Kumthong, N., *et al.* (2005) Cardanol polysulfide as a vulcanizing agent for natural rubber. *J. Sci. Res. Chula. Univ.*, **30** (1), 23–40.
66. Yadav, R. and Srivastava, D. (2008) Studies on cardanol-based epoxidized novolac resin and its blends. *Chemistry*, **2** (3), 173–184.

67. Philip, J.Y.N., Buchweishaija, J., Mkayula, L.L. and Ye, L. (2007) Preparation of molecularly imprinted polymers using anacardic acid monomers derived from cashew nut shell liquid. *J. Agric. Food Chem.*, **55**, 8870–8876.

68. Mmongoyo, J.A., Mgani, Q.A., Mdachi, S.J.M., *et al.* (2012) Synthesis of a kairomone and other chemicals from cardanol, a renewable resource. *Eur. J. Lipid. Sci. Technol.*, **114**, 1183–1192.

69. Baader, S., Podsiadly, P.E., Cole-Hamilton, D.J. and Goossen, L.J. (2014) Synthesis of tsetse fly attractants from a cashew nut shell extract by isomerising metathesis. *Green Chem.*, **16**, 4885–4890.

70. Mgaya, J.E., Mubofu, E.B., Mgani, Q.A., *et al.* (2015) Isomerization of anacardic acid; a possible route to the synthesis of an unsaturated benzolactone and a kairomone. *Eur. J. Lipid Sci. Technol.*, **117**, 190–199.

71. Julis, J., Bartlett, S.A., Baader, S., *et al.* (2014) Selective ethenolysis and oestrogenicity of compounds from cashew nut shell liquid. *Green Chem.*, **16**, 2846–2856.

72. Mubofu, E.B., Mdoe, J.E.G. and Kinunda, G. (2011) The activity of invertase immobilized on cashew nut shell liquid-templated large pore silica hybrids. *Catal. Sci. Technol.*, **1**, 1423–1431.

73. Hamad, F.B., Mubofu, E.B. and Makame, Y.M. (2011) Wet oxidation of maleic acid by copper(II) Schiff base catalysts prepared using cashew nut shell liquid templates. *Catal. Sci. Technol.*, **1**, 444–452.

74. Mlowe, S., Nejo, A.A., Rajasekhar Pullabhotla, V.S.R., *et al.* (2013) Lead chalcogenides stabilized by anacardic acid. *Mater. Sci. Semicond. Process.*, **16**, 263–268.

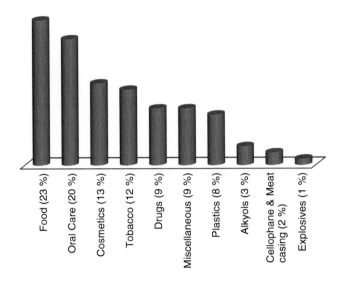

Figure 1.2 Commercial consumption of glycerol (industrial sectors and volumes).

Figure 1.3 Development of the model reaction in glycerol [32]: (a) beginning of the reaction as an identical mixture; (b) partial precipitation of the reaction; (c) the end of the precipitation of the reaction. From He *et al.* (2009) *Green Chem.*, **11**, 1767–1773. Reproduced by permission of RSC.

Bio-Based Solvents, First Edition. Edited by François Jérôme and Rafael Luque.
© 2017 John Wiley & Sons Ltd. Published 2017 by John Wiley & Sons Ltd.

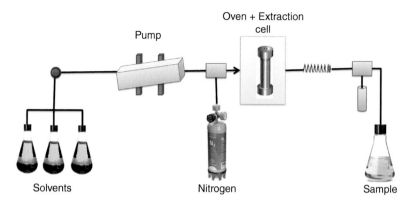

Figure 2.3 Accelerated solvent extraction.

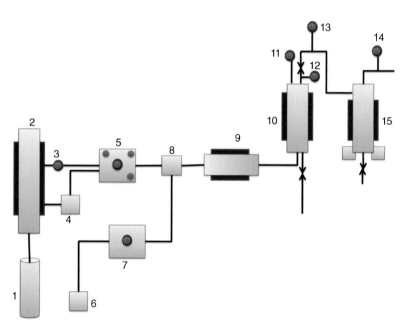

Figure 2.4 Schematic diagram of the supercritical extraction system: 1, CO_2 cylinder; 2, cooling heat exchanger; 3, flowmeter; 4, cooling bath; 5, CO_2 pump; 6, co-solvent reservoir; 7, co-solvent pump; 8, mixer; 9, heat exchanger; 10, extraction vessel; 11, pressure gauge; 12, temperature sensor; 13, automated back-pressure pump; 14, pressure gauge; 15, cyclone separator [22].

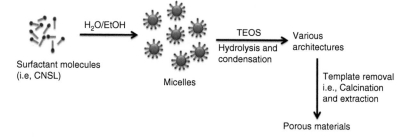

Figure 2.7 Preparation of porous materials employing CNSL as a templating agent.

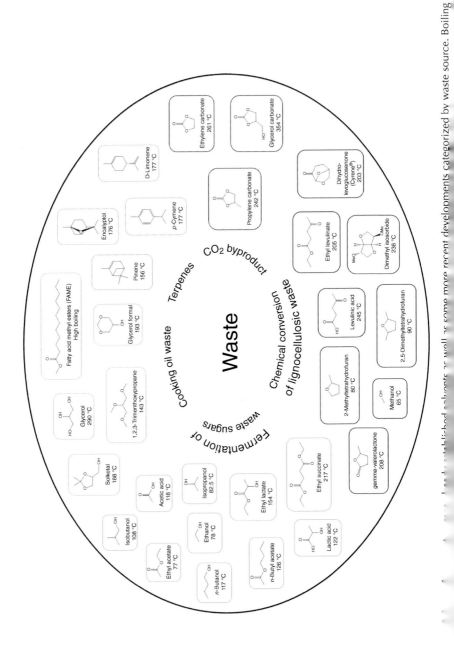

A circular diagram categorizing waste sources. Central label: **Waste**, with segments labeled: Terpenes, CO₂ byproduct, Chemical conversion of lignocellulosic waste, Fermentation of waste sugars, Cooking oil waste.

Compounds shown with boiling points:

Terpenes / Cooking oil waste section:
- D-Limonene 177 °C
- Eucalyptol 176 °C
- p-Cymene 177 °C
- Pinene 156 °C
- Fatty acid methyl esters (FAME) High boiling
- Glycerol formal 193 °C
- 1,2,3-Trimethoxypropane 143 °C
- Glycerol 290 °C
- Solketal 188 °C
- Isobutanol 108 °C
- Acetic acid 118 °C
- Ethyl acetate 77 °C
- n-Butanol 117 °C

CO₂ byproduct section:
- Ethylene carbonate 261 °C
- Glycerol carbonate 354 °C
- Propylene carbonate 242 °C

Chemical conversion of lignocellulosic waste section:
- Dihydro-levoglucosenone (Cyrene®) 203 °C
- Ethyl levulinate 205 °C
- Dimethyl isosorbide 238 °C
- Levulinic acid 245 °C
- 2,5-Dimethyltetrahydrofuran 90 °C
- 2-Methyltetrahydrofuran 80 °C
- Methanol 65 °C
- gamma-valerolactone 208 °C

Fermentation of waste sugars section:
- Ethyl succinate 217 °C
- Isopropanol 82.5 °C
- Ethyl lactate 154 °C
- Ethanol 78 °C
- Lactic acid 122 °C
- n-Butyl acetate 126 °C

...and published solvents as well as some more recent developments categorized by waste source. Boiling

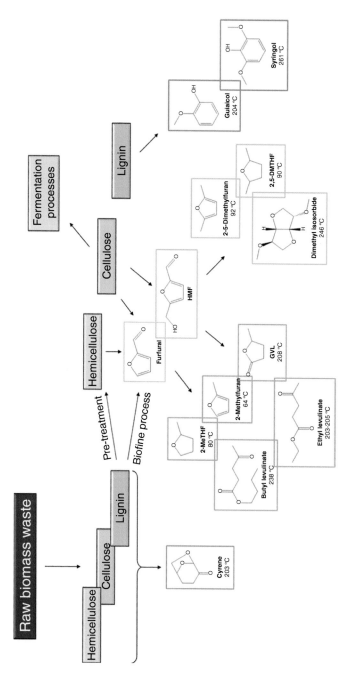

Figure 3.2 Lignocellulosic waste solvents produced by chemical transformations.

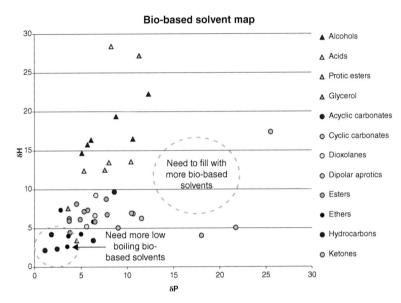

Figure 3.13 Map of current bio-based solvents using HSP hydrogen-bonding versus polarity scales. Current bio-based hydrocarbons (purple circles) are high-boiling. There is a need to develop lower-boiling bio-derived solvents as alternatives. There is also an empty space where dipolar aprotic solvents such as NMP, DMF and DMSO would be located, which must be filled with bio-based solvents.

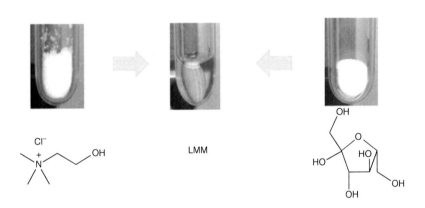

Scheme 4.10 Formation of an LMM between ChCl and fructose.

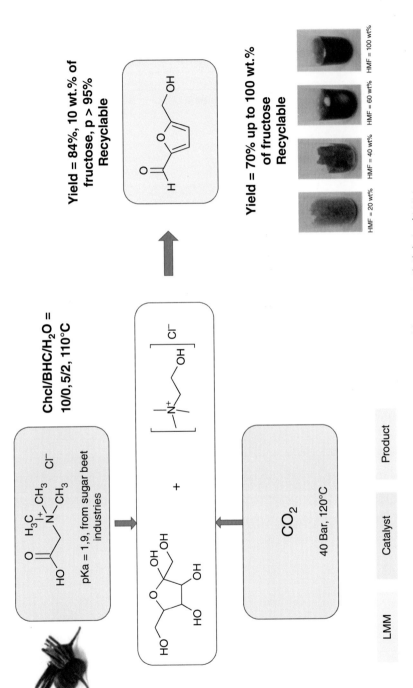

Scheme 4.13 Acid-catalysed dehydration of fructose in ChCl-derived LMM.

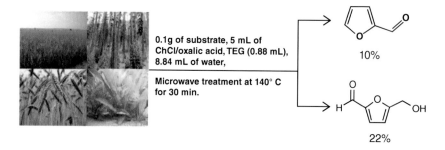

0.1g of substrate, 5 mL of
ChCl/oxalic acid, TEG (0.88 mL),
8.84 mL of water,

Microwave treatment at 140° C
for 30 min.

10%

22%

Scheme 4.16 Conversion of lignocellulosic biomass to HMF and furfural in DES.

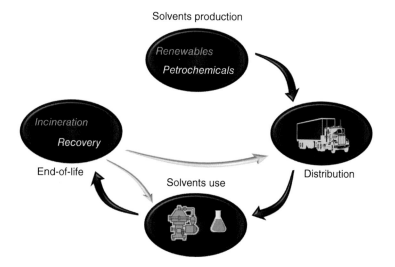

Figure 6.1 The life cycle of a solvent.

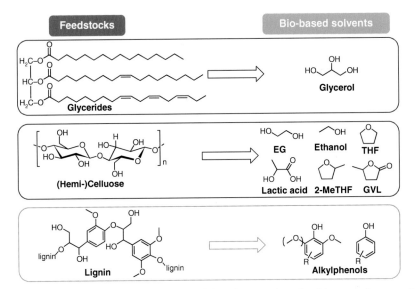

Figure 7.1 Bio-based solvents manufactured from biomass by using bio- and chemocatalysis. EG, ethylene glycol; THF, tetrahydrofuran; 2-MeTHF, 2-methyltetrahydrofuran; GVL, gamma-valerolactone.

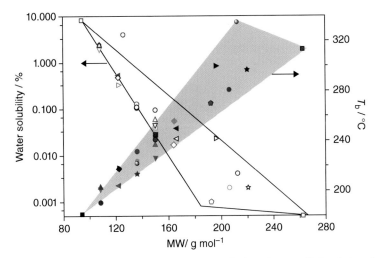

Figure 7.2 The water solubility and boiling point (T_b) of selected mono-alkylated phenols as a function of molecular weight (MW) [20–32]. The hollow symbols represent the water solubility and the filled symbols represent the boiling points of selected mono-alkylated phenols at 25°C.

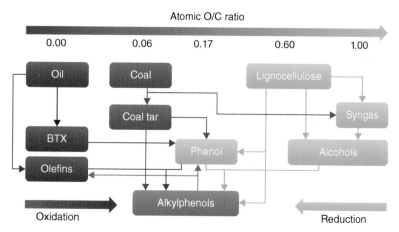

Figure 7.3 Methods to produce alkylphenols from different substrates. BTX, benzene, toluene and xylene.

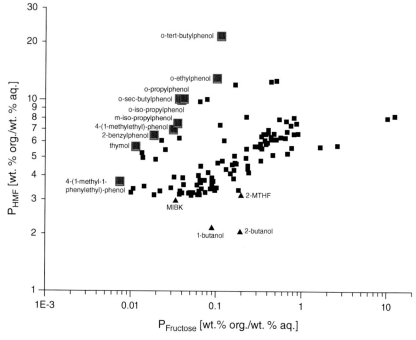

Figure 7.8 The 110 feasible solvents for extracting HMF from aqueous phase [139]. Each point represents a solvent: the triangles are the solvents used for verification of the computational method (COSMO-RS); the squares are the solvents identified through COSMO-RS to show improved extraction performance. The larger marked square points are alkylphenols. Adapted from Blumenthal *et al.* (2016) *ACS Sustain. Chem. Eng.*, **4**, 228–235. Reproduced with permission of the American Chemical Society.

3

Solvents from Waste

Fergal Byrne, Saimeng Jin, James Sherwood, C. Rob McElroy, Thomas J. Farmer, James H. Clark, and Andrew J. Hunt
Green Chemistry Centre of Excellence, Department of Chemistry, University of York, York, UK

3.1 Introduction

Solvents should be sustainable from an economic, social and environmental perspective so that the extensive supply chains (and therefore the many jobs associated with the solvent sector) are protected and market growth is harmonious with global ambitions of environmental protection [1]. Bio-based solvents must overcome two major obstacles to be considered as sustainable. Firstly, they must be able to compete with traditional solvents with respect to the cost of their production. At present, bio-based products tend to be produced at relatively small scales and as a consequence are more expensive than equivalent petrochemical products [2]. A green premium can be justified in some cases, but solvents, being low-value chemicals, are particularly vulnerable in a price-competitive market. It can be anticipated that a combination of the declining supply of accessible petroleum sources and the ever-improving efficiency and scale of chemical production from biomass will encourage more economically competitive bio-based products. Secondly, there are still concerns about land use for non-food crops. Land that is currently used for food production should not be reassigned to biofuel or chemical production. Even

Bio-Based Solvents, First Edition. Edited by François Jérôme and Rafael Luque.
© 2017 John Wiley & Sons Ltd. Published 2017 by John Wiley & Sons Ltd.

if the scale and growth of food production are undisturbed, this debate has since evolved into the more complex issue of 'indirect land use change' [3]. When agricultural crops are grown on land previously covered by forest or other high-carbon stock areas, an increase in greenhouse gas emissions counteracts the benefit of producing bio-based products to replace petrochemicals. It follows therefore that the by-products of food production and industry, and more broadly the waste generated along the food supply chain (FSC), are an ideal resource for solvent production. This diverse and abundant source of chemicals, if exploited correctly, permits food, chemical and fuel production with no major reassigning of land required.

Waste is available from many sources and can be categorized into four main groups: industrial, agricultural, sanitary and solid urban waste. Food waste occurs as both agricultural waste (upstream in the FSC) and solid urban residue groups (downstream in the FSC) [4]. About 89 million tonnes of edible food waste were produced in 2006 in the EU 27 as estimated by the European Commission [5], and the UN estimates the global figure as 1.3 billion tonnes [6]. Exploitation of this resource for the production of chemicals and fuels is one promising route to sustainable chemical production [7] (Figure 3.1).

Food waste can be further subdivided into pre-consumer and post-consumer categories [4]. While post-consumer food waste consists of a wide range of mixed biomass, demanding intensive pre-treatment, pre-consumer food waste is a more attractive feed for the production of chemicals, as it usually comprises one type of biomass, simplifying pre-processing and separation. Pre-consumer food waste consists of agricultural waste (rice husks, wheat straw, etc.), food processing waste (tomato pomace, apple pomace, grape pomace, etc.) and domestic food waste packaging and distribution (damaged goods, expired fresh foods, etc.), which can be used as the source of a variety of bio-platform molecules for the production of solvents. Much research has been carried out on the valorization of different varieties of food waste, as highlighted in a recent review [4]. One prominent example is citrus peel, with over 15 billion tonnes of citrus waste generated each year. It contains extractible D-limonene, which is itself a useful solvent and can be further modified into other solvents, including *p*-cymene for example [8]. Like other food wastes such as grape pomace, wheat straw, corn stover, nut shells and sugarcane bagasse, the structural basis of this biomass is lignocellulose. The major cellulose portion is particularly suitable for conversion into bio-platform molecules by both fermentation and thermochemical means, and then on to solvents. Lignin is harder to exploit as a chemical feedstock, but some examples of lignin-derived solvents now exist [9].

Forestry waste, an industrial by-product that includes waste wood, bark and other plant matter, is composed of the same lignocellulosic biomass as food waste and can be processed in much the same way. Carbon dioxide can be thought of as another industrial waste product, created from fermentation and combustion [10]. Carbon dioxide can be used in its supercritical and near-critical forms [11], and reacting carbon dioxide with other molecules obtained from waste sources

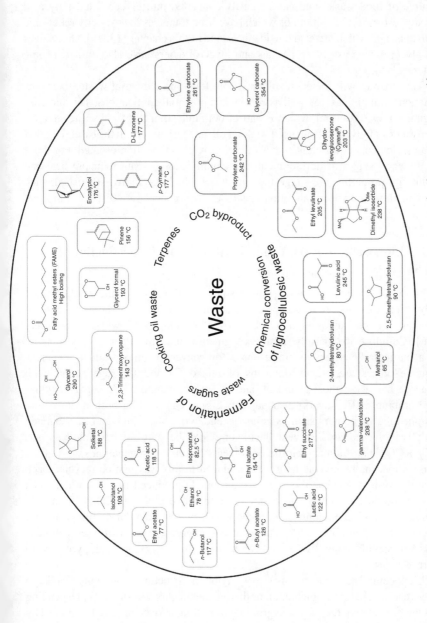

Figure 3.1 Solvents from waste: some already established solvents as well as some more recent developments categorized by waste source. Boiling points have been given for each solvent. (A colour version of this figure appears in the plate section.)

can yield other useful solvents, notably cyclic carbonates [12], which are outside the scope of this chapter. Waste cooking oil from the catering industry is another source of food waste, and can be easily converted into glycerol and fatty acid methyl esters (FAME), both of which have been touted as low-toxicity renewable solvents, and which are also outside the scope of this chapter [13, 14]. Derivatives of glycerol such as glycerol ketals and glycerol ethers have also attracted some interest [15].

This chapter will focus on solvents from food and industrial waste, their production from biomass, as well as current and potential applications. This includes lignocellulosic material, which is part of both industrial and food waste streams, derivatives of glycerol from used cooking oil, and terpenes from the essential oil of plants. Solvents at different stages of commercialization are presented. For those at the early stages of development, some applications to suit their physical and solubility properties are suggested, and those which are more established, such as 2-methyltetrahydrofuran, ethyl lactate and D-limonene, are discussed in more detail.

3.2   Lignocellulosic Waste as a Feedstock for the Production of Solvents

Lignocellulosic waste is an appealing source of biomass for the production of bio-based solvents. It is composed of lignin, cellulose and hemicellulose in amounts in the ranges of about 10–35, 25–60 and 25–40 wt.%, respectively [16]. Cellulose exclusively comprises glucose units in linear chains, making it a particularly attractive bio-resource for conversion into products, whereas hemicelluloses are branched chains of mixed C_5 and C_6 sugar units, with glucose being the most abundant, and galactose, mannose (C_6 sugars), xylose and arabinose (C_5 sugars) also present. Lignin can also be broken down into its phenolic monomers such as guaiacols and syringols, which can be separated or possibly used as a solvent blend; or it can be pyrolysed to make syngas ($H_2 + CO$) [17]. As many bio-based solvents require hydrogenation steps in their production, the renewable hydrogen produced in this way can be recycled back into the system. Methanol, which is traditionally produced from petroleum, can also be produced from syngas by the reaction between H_2 and CO [18]. Varying amount of insoluble humins are produced as a side product during lignocellulose processing in variable yields. This char is composed of unhydrolysed cellulose and hemicellulose and lignin, and can be burnt for energy or again pyrolysed to form syngas.

Lignocellulosic biomass is very rigid in structure and must undergo pre-treatment before fermentation or thermochemical processing [19]. Depending on the biomass source, physical grinding or milling may be required, followed by chemical, physicochemical or biological treatment. Transformations to bio-based solvents can be carried out by several approaches. The first route is to use ground or milled lignocellulosic material directly to produce bio-platform molecules such as levulinic acid, hydroxymethylfurfural (HMF) and furfural, which can then be converted to solvents (e.g. the Biofine process) [20]. The bio-based solvent dihydrolevoglucosenone (cyrene) can also be made at this stage [21]. Another option is to free the sugar units from purified cellulose by means of chemical, physicochemical or biological treatments, which are then converted to platform molecules by either chemical transformations or fermentation.

Chemical treatment to separate the lignocellulosic components includes dilute acid treatment (using mineral acids), organosolv treatment (using an organic solvent such as ethanol to dissolve hemicellulose and lignin), or alkali treatment (using hydroxides or lime) and aim to yield pure cellulose [22]. These products can all be converted to solvents by further modification. Ionic liquids can be used to dissolve different fractions of lignocellulose, although difficulties may arise during subsequent separation and fermentation [23–25]. Additionally, the 'greenness' of many ionic liquids is questionable [26].

Physicochemical methods involve the use of a physical effect in combination with a catalyst such as sulfuric acid [27]: for example, steam explosion, which as the name suggests uses steam to break up the cell structure; hydrothermolysis, which is similar to steam explosion but uses liquid hot water instead of steam; wet oxidation treatment, which uses hot water and pressurized air; or ammonia fibre explosion (AFEX), which uses ammonia in a similar manner to steam explosion.

Biological treatment can also be applied, which involves the use of different kinds of fungi to degrade lignocellulose. White rot fungi has been found to be particularly useful, as it specifically degrades lignin, releasing cellulose [28].

To obtain the free sugar units from cellulose, enzymatic or acid hydrolysis is carried out. Following this, two main types of processing can be applied to yield valuable chemicals, namely chemical transformations and fermentation, and their products will be described below, as well as the production of cyrene directly from lignocellulose.

3.2.1   Chemical Transformations of Sugars

Upon acid treatment of sugars, two platforms arise from which solvents can be synthesized depending on the raw material input: C_5 sugars will form furfural and C_6 sugars can form both HMF and furfural depending on the process conditions (Figure 3.2). Both are important bio-platform molecules that can be sourced from waste [29, 30], and many potential bio-based solvents can be produced from them by catalytic treatment, such as hydrogenolysis, hydrogenation, dehydration and esterification.

2-Methyltetrahydrofuran (2-MeTHF) from the furfural platform is an excellent bio-based solvent due to its wide range of solvation and physical properties. It is an aprotic Lewis base with a boiling point of 80°C with a polarity and basicity between that of diethyl ether and tetrahydrofuran (THF) [29]. It can potentially

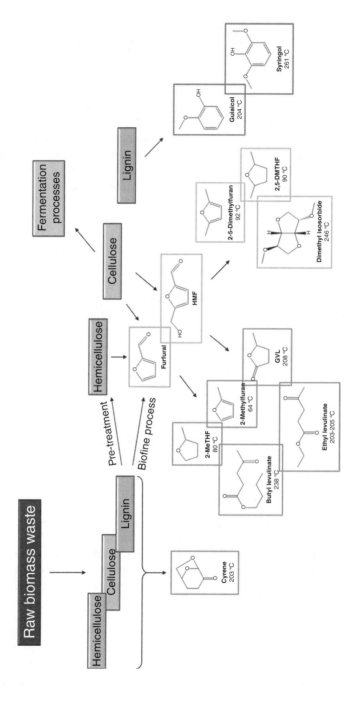

Figure 3.2 Lignocellulosic waste solvents produced by chemical transformations. (A colour version of this figure appears in the plate section.)

replace many common solvents such as THF, toluene, dichloromethane (DCM) and diethyl ether in certain applications. 2-MeTHF is at an advanced stage of commercial development, and is currently produced at a large scale exclusively from biomass. There are two main routes to its production. It can be produced from furfural by hydrogenation to furfuryl alcohol and then hydrogenation to methyl furan using metal catalysts such as palladium or ruthenium (Figure 3.3, route 1). Alternatively, it can be produced by hydration of HMF to form levulinic acid, with formic acid being produced as a by-product (Figure 3.3, route 2). An advantage of this second route is that expensive and rare catalysts are not required. Instead, a nanocomposite copper/silica catalyst can be used to reduce and cyclize levulinic acid to 2-MeTHF via γ-valerolactone (GVL) with almost complete selectivity. Increasing the copper loading in the catalyst favours 2-MeTHF formation with good yields [31].

Preliminary toxicological investigations suggest that exposure to 2-MeTHF is not linked to mutagenicity or genotoxicity [32]. One application in which 2-MeTHF excels is in organometallic chemistry, where it can directly replace THF. It is less miscible in water than THF and has the rare property of becoming less water miscible at higher temperatures [33]. It also forms a favourable azeotrope with water, making drying easier. This is a particularly valuable property for use in Grignard [34], Reformatsky [35], lithiation [36] and metal-catalysed coupling [37] reactions, and for the work-up of amide coupling reactions [38].

Milton and Clarke [39] demonstrated that the Kumada cross-coupling of Grignard reagents with a range of aryl halides can be performed in 2-MeTHF while dramatically decreasing the use of the solvent in comparison with the traditional process. In the same paper, 2-MeTHF was employed in the synthesis of the $Pd(L)Cl_2$ pre-catalysts, where L is a bidentate phosphine ligand, starting from tetrachloropalladate and using microwave heating. Kadam *et al.* [34] showed that 2-MeTHF outperformed or equalled the performance of THF and diethyl ether in the preparation of many Grignard reagents as well as in the follow-up reactions. Recently, Smoleń *et al.* [40] found that the traditional solvents for ruthenium-catalysed olefin metathesis such as DCM and toluene can be substituted with 2-MeTHF with equal or comparable results.

2-MeTHF can also affect chemoselectivity and enantioselectivity due to its stereocentre at the 2-position. Zhong *et al.* [41] disclosed that 2-MeTHF gave higher chemoselectivities and higher yields in Grignard reactions than THF, with the additional benefit that the solvent could be subsequently recycled and reused. Robert *et al.* [42] found that 2-MeTHF can dramatically improve the enantioselectivity for the reaction of Cu-catalysed 1,4-addition of Grignard reagents to cyclohexenone. In the asymmetric reduction reaction of acetophenone with $(EtO)_2MeSiH$ catalysed using iridium complex and $AgBF_4$, Kawabata *et al.* [43] reported that, when 2-MeTHF was used as a solvent in the reaction, a high enantioselectivity of 92% with high yield of 71% was obtained.

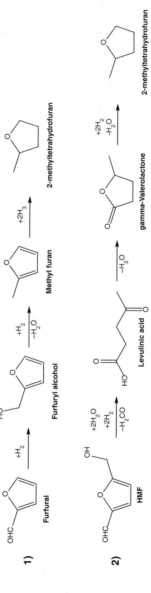

Figure 3.3 The production of the bio-based solvent 2-MeTHF from furfural.

Owing to the low miscibility of 2-MeTHF with water, it can be used as a solvent for the extraction of organic compounds from the aqueous phase. Vom Stein *et al.* [44] synthesized furfural via dehydration of xylose in a biphasic system using $FeCl_3 \cdot 6H_2O$ and NaCl in an aqueous phase and 2-MeTHF as the organic phase. In this way 98% of the furfural was extracted into the 2-MeTHF phase [44].

2-MeTHF has also found applications as a solvent in biochemical transformations. Chen *et al.* [45] reported that 2-MeTHF was a suitable reaction solvent for immobilized *Candida antarctica* lipase B (Novozym 435). Duan and Hu [46] demonstrated that 2-MeTHF was used effectively as a reaction solvent for enzyme-mediated transphosphatidylation of phosphatidylcholine with L-serine for the production of phosphatidylserine. Shanmuganathan *et al.* [47] discovered that 2-MeTHF is a potential replacement for dimethylsulfoxide (DMSO) as a co-solvent or methyl *tert*-butyl ether (MTBE) as a second organic phase in lyase-catalysed reactions.

Although 2-MeTHF exhibits many superior properties to THF in significant applications, there are still some problems with its use. The ether group in 2-MeTHF leads to the formation of peroxide compounds during storage, which have a high risk of explosion, meaning that antioxidants are required. Additionally, the price of 2-MeTHF is currently higher than that of many traditional solvents that it could replace, although it is assumed that, with increasing demand, its price will drop. In spite of these problems, 2-MeTHF is still a valuable solvent due to its overall environmental impact and its origin from renewable materials.

2,5-Dimethyltetrahydrofuran (2,5-DMTHF) is a structural cousin of 2-MeTHF, but from the waste-derived HMF platform. Although less established than 2-MeTHF, it could be used for many similar applications. Interestingly, it can theoretically be produced more economically, as the HMF platform utilizes the more abundant C_6 sugars in cellulose, although at present less research has been carried out for its use as a solvent. It can potentially be used for many of the same reactions as 2-MeTHF and in fact its performance may even be improved due to its higher partition coefficient, making drying easier. 2,5-DMTHF has a boiling point of 90°C and is of lower overall polarity than 2-MeTHF and THF, but, like many other ethers, it is susceptible to peroxide formation, so preventive measures must be taken during its use and storage. Grochowski *et al.* [48] reported a neat, one-step synthesis from waste corn stover that utilized a dual catalytic system consisting of a rhodium catalyst and hydrogen iodide.

2-Methylfuran and 2,5-dimethylfuran are the precursors to 2-MeTHF and 2,5-DMTHF, which have received less attention for use as solvents, but instead have mainly been touted as bio-based fuel additives. However, they too are aprotic molecules of medium polarity, which could be useful in similar applications to THF. They have the advantage over THF, 2-MeTHF and 2,5-DMTHF in that they are more resistant to auto-oxidation and so are safer to use [49], although their chemical stability may be an issue in some reaction conditions.

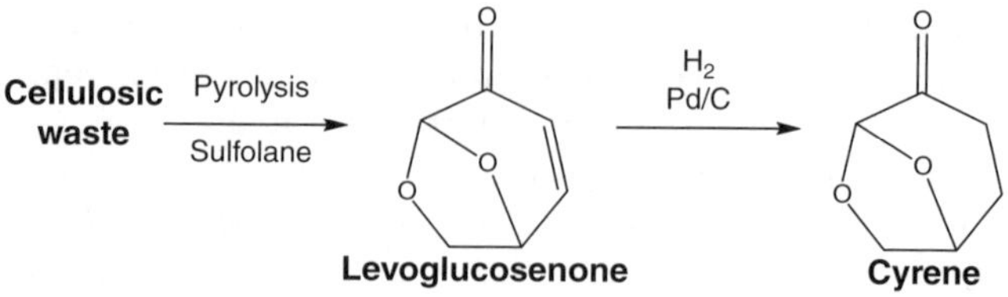

Figure 3.4   The synthesis pathway of cyrene from cellulose waste.

Sherwood *et al.* [21] have developed a new solvent, dihydrolevoglucosenone (cyrene), in collaboration with Circa. Cyrene can be synthesized from cellulose waste by a two-step reaction as illustrated in Figure 3.4 [50]. The second step is a solvent-free hydrogenation under high pressures with no loss of selectivity in comparison to when a solvent is used.

Cyrene is a polar aprotic solvent whose polarity is similar to that of *N*-methyl pyrrolidone (NMP), *N,N*-dimethyl formamide (DMF) and sulfolane. It is a chiral, multifunctional molecule made up of two fused rings, which lie almost perpendicular to each other, making cyrene quite different from conventional solvents. The application of cyrene in some preliminary examples of reaction chemistry has shown comparable results to NMP and other reprotoxic dipolar aprotic solvents [21]. At present, studies are limited and so the extent to which cyrene can substitute for NMP and similar solvents is not yet fully understood. With reactive carbonyl functionality and the potential to decompose at high temperatures, the utility of cyrene will be restricted to a limited number of applications. Users of solvents should be prepared to accept that, to achieve the substitution of toxic and unsustainable solvents, several solvents may be required to cover the different applications presently only needing one solvent. This situation is well illustrated by the presence of many new, greener options now available for substitution in the polar aprotic solvent space (cyrene, GVL, cyclic carbonates, etc.).

Gamma-valerolactone (GVL) has many appealing properties that make it an alternative bio-based polar aprotic solvent. The high polarity, low vapour pressure, low toxicity and low peroxide-forming potential of GVL are all beneficial attributes [51]. It is similar in polarity to *N,N*-dimethyl acetamide (DMAc) and NMP as measured by Hansen solubility parameters, and, given the chronic toxicity of amides, GVL could be an ideal replacement for these solvents (Table 3.1). It is produced from either the HMF or furfural platforms: HMF can be hydrated to form levulinic acid or furfural can be hydrogenated to furfuryl alcohol, which in turn is hydrolysed to form levulinic acid [20, 52]. Hydrogenation and subsequent dehydration of levulinic acid yields GVL (Figure 3.5) [53].

Qi and Horváth [54] reported that GVL can be applied as a solvent in its own synthesis from fructose in a one-pot reaction, thus dramatically reducing work-up and isolation. Similarly GVL has also found use as a solvent for the production

Table 3.1 Comparison of the Hansen solubility properties of GVL, DMAc and NMP, which were generated by the HSPiP program

Solvent	δ_D	δ_H	δ_P
GVL	16.9	6.3	11.5
DMAc	16.8	9.4	11.5
NMP	18	17.2	12.3

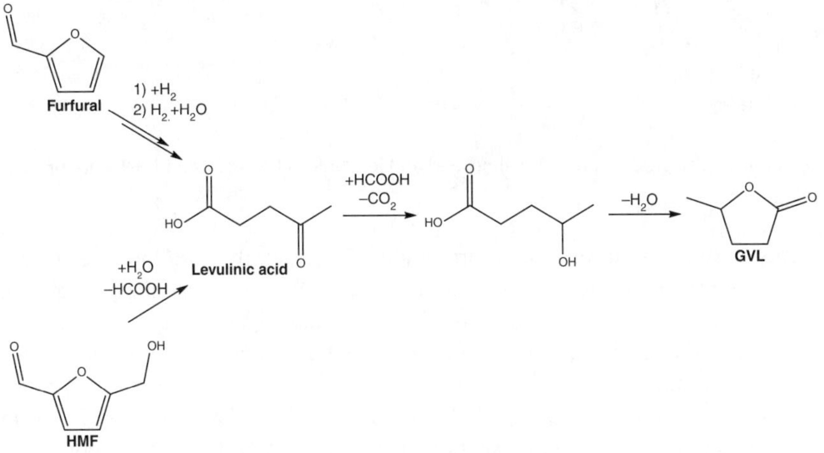

Figure 3.5 Synthetic routes to GVL from the HMF and furfural platforms.

of the platform molecules HMF, furfural and levulinic acid from sugars [55]. Additionally, it has found use as a dipolar aprotic solvent for Heck [56] and Sonogashira coupling reactions [57], with comparable results to traditional dipolar aprotic solvents in all cases (albeit with longer reaction times). Excellent results were also obtained for the Hiyama coupling reaction in GVL [58]. Qi et al. [59] successfully reacted fructose, glucose or sucrose to give HMF or levulinic and formic acid at good selectivity through manipulation of conditions using H_2SO_4 as catalyst and GVL as solvent. Gallo et al. [60] discovered that HMF can be synthesized from fructose and glucose with high yields (80% from fructose and 63% from glucose) by employing GVL as solvent. Alonso et al. [55] also found that the cellulosic fraction and hemicellulose fraction of waste lignocellulosic biomass (corn stover) can be converted into levulinic acid and furfural respectively under the same conditions by using GVL as a solvent.

Levulinic acid and its esters such as methyl, ethyl and butyl levulinate also have potential uses as bio-based solvents and some have already seen commercialization [61]. All are high-boiling (>200°C), medium-polarity molecules that have the solubility properties to replace hazardous solvents such as DCM in cleaning

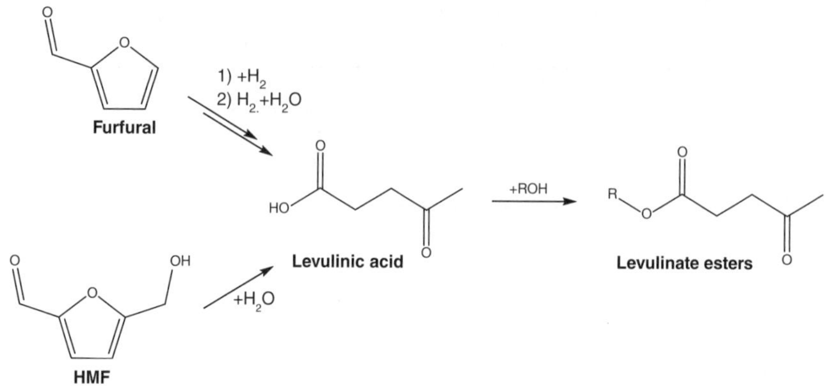

Figure 3.6 The production of levulinic acid and levulinic acid esters from HMF and furfural.

agents and paint removers. They are produced very easily from biomass in several ways: direct esterification of levulinic acid using the corresponding alcohol in acidic conditions; via the alcoholysis of furfuryl alcohol; or via the dehydration of levulinic acid to angelica lactone and subsequent esterification to yield levulinate (Figure 3.6) [55].

Dimethyl isosorbide is the methyl ether of isosorbide that can be produced in basic conditions from sorbitol using dimethyl carbonate as the methyl source and solvent (Figure 3.7) [62]. As a solvent, it is aprotic and of medium polarity, and is of low volatility. Isosorbide is a promising basis for a number of bio-based surfactants [63, 64]. Broader interest in isosorbide may be helpful in encouraging larger-scale production and lower costs, thus opening up the possibility for isosorbide ethers as viable solvent products. A recent application of dimethyl isosorbide is epoxidation chemistry, although results were far superior in 2-MeTHF [65]. The solvent system of dimethyl isosorbide used as a co-solvent with water has been found to be suitable for the Baylis Hillman reaction. Better results were seen compared to when either water or dimethyl isosorbide are used individually as solvents. It also outperformed dioxane and an acetonitrile–water mixture [66].

3.2.2 Fermentation of Lignocellulosic Waste

With ever-improving chemical engineering of microorganisms, it is becoming possible to generate a wider range of platform molecules with satisfactory selectivity than achievable through conventional fermentation [67–70]. The alcohols ethanol, isopropanol, *n*-butanol and isobutanol, the diols 1,3-propanediol and 2,3-butanediol, and the acids acetic acid, propionic acid, lactic acid and succinic acid have all been produced from waste biomass using fermentation processes [71]. Aside from the solid succinic acid and gaseous ethene and isobutene, all

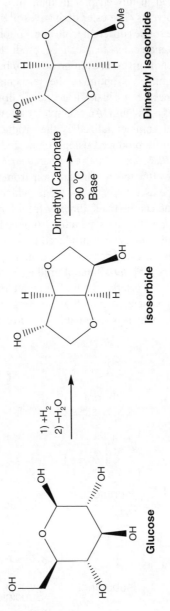

Figure 3.7 Synthesis of dimethyl isosorbide from glucose.

can be used as solvents directly or can undergo some chemical manipulation to generate different solvents. Ethanol and the low-molecular-weight alcohols and acids will not be discussed in this chapter in their own right because of their well-established use as solvents. The ester derivatives of succinic acid and lactic acid provide valuable case studies different from conventional solvents. Figure 3.8 shows a range of small molecules that can be produced from fermentation. Combining acids and alcohols gives an even greater number of ester solvents.

It is important to continue to optimize the fermentation of biomass and subsequent separation processes to make it economically competitive with the cracking and processing of petroleum. An example of a new fermentation process that has been refined and commercialized is the production of isobutanol by Gevo [72]. Isobutanol can be produced using engineered microorganisms such as *Escherichia coli*, *Basillus subtilis* and *Corynebacterium glutamicum*. While isobutanol itself does not offer too much variation from 1-butanol in terms of solvent properties, it can be used to make *p* -xylene, which can then also be used as a solvent [73]. Isobutene can be made directly from fermentation using an *E. coli* strain with the advantage of easier separation from the fermentation broth due it being a gas [74]. Recently, it has been produced by fermentation from xylose, which is a major component of wood hemicellulose, giving it the advantage of not competing with food sugars [75]. *tert*-Butanol produced from bio-isobutene [76] can be esterified into *tert*-butyl esters, which offer an alternative to *n*-butyl esters [77]. *tert*-Butyl esters are characteristically lower boiling and less polar than their straight-chain counterparts, which is a valuable property, as there are a limited

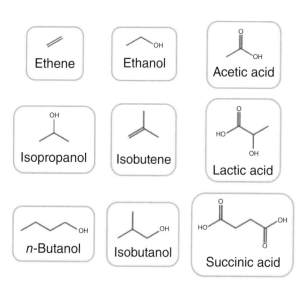

Figure 3.8 Low-molecular-weight alcohols, acids and olefins that can be produced by fermentation of waste cellulose and which can be further modified to make solvents.

number of bio-based solvents to replace low-polarity, low-boiling hydrocarbons. An example is *tert*-butyl acetate, which has a boiling point range of 94–96°C. It is not classified as a volatile organic compound (VOC) by the US Environmental Protection Agency (EPA) and it has been designated as a green solvent in the GlaxoSmithKline solvent selection guide [78, 79].

Actinobacillus succinogenes has been shown to successfully ferment various waste streams to produce the bio-platform molecule succinic acid. This includes both five- and six-membered sugars from corn straw waste [80], bread waste [81], which gave rise to the highest yields of succinic acid compared to other food waste sources (0.55 g per gram of bread), mixed orange peel and wheat straw. However, the citrus waste required pre-treatment to remove D-limonene due to even low concentrations exhibiting an inhibitory effect on the *A. succinogenes* S85 strain [82]. Additionally, mixed food waste has been exploited using recombinant *E. coli* to produce succinic acid without the formation of any additional by-products [83].

Lactic acid is another potential solvent that can be sourced by fermentation of sugars. Although at present it is mainly produced from corn starch or sugar [84, 85], a number of industrial producers of lactic acid aim to move to, or are already exploiting, non-food sugar sources such as food and agricultural waste [86]. There are several examples of its production from waste sources already in the literature: potato peelings [87], mixed fruit and vegetables [88], municipal solid waste [89] and mixed food waste [90]. Of particular interest, Liang *et al.* [87] produced lactic acid from potato peel waste using a mixed, undefined microbial culture in unsterile conditions with blending as the only pre-treatment required to give moderate yields of 0.22 g lactic acid per gram of potato peel. However, when compared to many other processes for the production of lactic acid, these lower yields could be compensated by the energy and materials savings due to the absence of sterilizing and pre-treatment. Li et al. [90] produced high yields of optically pure L-lactic acid from mixed food waste by the inhibition of D-lactic acid-producing enzymes using sewage sludge and intermittent alkaline fermentation [90].

Current separation methods of lactic acid from the broth is a major contributor to the cost of its production. Traditionally, the broth is neutralized with calcium carbonate, followed by several separation steps before the lactic acid is freed from the calcium lactate using sulfuric acid. This requires the use of large amounts of sulfuric acid and produces calcium sulfate as a by-product [91]. New methods are being investigated, such as reactive distillation, in which the acid is esterified using either ethanol or methanol [92]. However, distillation is difficult due to the high boiling points of the mixture components. Recently, nanofiltration coupled with electrodialysis has also been shown to be an effective method of purifying lactic acid to a high degree, with 85.6% selectivity for the L-isomer [93]. This is achieved in just two steps from the fermentation broth compared to multiple purification steps in other methods, which could compensate for any extra energy required.

Lactic acid can be used directly as a highly protic solvent and has been successfully applied in several synthetic organic chemistry applications. It is an alpha-hydroxy acid with a boiling point of 122°C that is immiscible with low-polarity solvents such as hydrocarbons or diethyl ether. Yang *et al.* [94] used lactic acid as solvent for three-component reactions between styrenes, formaldehyde and an active phenolic compound or *N,N*-dialkylacetoacetamides. Lactic acid gave improved yields compared to acetic acid in a range of cascade Knoevenagel/*oxo*-Diels–Alder type reactions, possibly due to the higher acidity of lactic acid. Multi-component reactions involving diethyl acetylenedicarboxylate, benzaldehydes and anilines, which previously used ethanol in combination with *p*-toluene sulfonic acid or tin chloride, were also successful, with shorter reaction times, and in some cases the product precipitated out of solution, easing separation. Similarly aniline-catalysed condensations between salicylaldehydes and diethyl acetylenedicarboxylate produced excellent yields, and a Friedländer annulation between 2′-aminoacetophenone and 1,3-dicarbonyl compounds provided product yields up to 95%. Clark *et al.* [95] have used lactic acid in the Biginelli reaction between urea, benzaldehyde and methyl acetoacetate, where it enhanced the Lewis acid-catalysed reaction compared to non-acidic solvents.

Lactic acid can be esterified with bio-ethanol to produce ethyl lactate [96]. It is a high-boiling (154°C), water-miscible solvent with low surface tension [96]. The US Food and Drug Administration (FDA) has permitted ethyl lactate to be used in food products due to its low toxicity [96]. As such, ethyl lactate is a component in a number of recently developed cleaning formulations [97]. It exhibits many other green properties: it is biodegradable, has low vapour pressure, is non-flammable, non-ozone-depleting and easily recyclable [98–100]. Ethyl lactate has also been successfully demonstrated as a substitute for a wide range of conventional environmentally hazardous or toxic solvents such as DCM, ethylene glycol ethers and chloroform [101].

Many applications of ethyl lactate as a solvent can be found in the literature. It has already been used in cycloadditions [100], amidations [102] and coating applications [103]. Paul and Das [104] developed an environmentally benign process for the synthesis of indenodihydropyridine and dihydropyridine derivatives by using ethyl-L-lactate as a solvent and (±)-lactic acid as the catalyst [104]. Wan *et al.* [105] developed a new method for the Suzuki–Miyaura reaction by using ethyl lactate as a green co-solvent with water. In this process, the catalyst system comprises water–ethyl lactate, Pd(OAC)$_2$ and K$_2$CO$_3$, while employing different aryl bromides and iodides to generate arylboronic acids under ligand-free conditions. Ethyl lactate exhibited a much better performance than many traditional polar organic solvents such as DMSO, 1,4-dioxane and DMF.

Dandia *et al.* [100] used ethyl lactate as a solvent for the synthesis of spiro-oxindole derivatives in a 1,3-dipolar cycloaddition reaction, giving high yields at room temperature. Xu *et al.* [106] employed aldehydes, thiourea and 1,3-dicarbonyl compounds to run Biginelli reactions in ethyl lactate, catalysed by trimethylsilyl chloride to produce a class of 3,4-dihydropyrimidinthiones in high yields. Liu *et al.* [107] found ethyl lactate to be a highly efficient solvent for the promotion of the oxidative coupling reaction of thiols without using any catalyst or additives, producing a wide range of disulfides. Salerno and Domingo [108] used ethyl lactate as a non-toxic medium for the manufacture of polycaprolactone particles by a thermally induced phase isolation technique. Stefanidis *et al.* [109] employed ethyl lactate as a solvent for the homogenization of bio-oil. Mondal *et al.* [110] produced HMF from bio-based sugars by the dehydration of fructose. Graphene oxide was used as the catalyst and choline chloride as an additive, with ethyl lactate as the reaction medium. Importantly, ethyl lactate was recycled after the recovery of both products and efficiently reused for the follow-up synthesis of the two compounds with high purity.

Dropping prices have resulted in greater uptake of ethyl lactate as a solvent for cleaning, replacing solvents such as methyl ethyl ketone and dichloromethane and as discussed above. This success story demonstrates the potential for the production of competitive, high-performance solvents from lignocellulosic waste streams. Many of the platform molecules produced by fermentation or chemical transformation of lignocellulose have other applications in the plastics sector, not just as solvents [85]. As is true of lactic acid and its derivatives, this helps to secure a market, justify a large scale of production, and ultimately will reduce costs.

3.3 Solvents from Used Cooking Oil

Over 2 million tonnes of glycerol were produced globally in 2012 [111], making it an abundant raw material for the production of solvents. It is a by-product of the transesterification of vegetable oil, including used cooking oil, with methanol to yield fatty acid methyl esters (FAME), commonly known as biodiesel [112]. Both FAME and glycerol can be used as bio-based solvents in their own right. However, this chapter will focus solely on derivatives of the latter: glycerol formal, solketal and 1,2,3-trimethoxypropane (Figure 3.9).

Glycerol formal exists as two species, a dioxane (six-membered ring) and a dioxolane (five-membered ring), which are formed by the acid-catalysed condensation of glycerol with formaldehyde [113]. It is a clear, viscous, low-volatility solvent with a high boiling point. It is of medium to high polarity with the hydrogen-bonding capability of common alcohols, and is stable to auto-oxidation. However, the dioxolane structural isomer in particular is susceptible to hydrolysis in acidic conditions [15]. Owing to the similar physical and chemical properties of the dioxane and dioxolane, their use as a solvent blend for formulations may be the best option, as separation would be inefficient and energy-intensive, with little advantage gained [114, 115]. Glycerol formal has also been reported as a replacement for methyl isobutyl ketone for the preparation of the active pharmaceutical ingredient Loperamide, chosen for its low toxicity, high auto-ignition temperature and high flash point [116].

Solketal is another glycerol-based solvent formed by an acid-catalysed reaction, this time with acetone. Like glycerol formal, solketal exists as both a

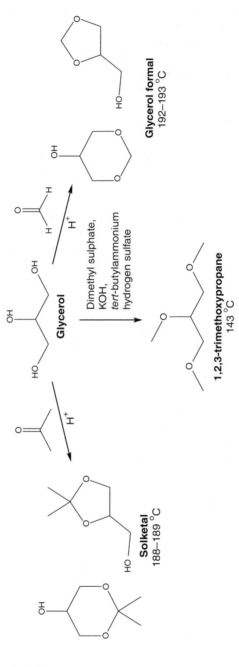

Figure 3.9 Potential solvents derived from glycerol, with their corresponding boiling points.

six-membered ring and a five-membered ring, is of low toxicity and has a similar polarity to ethyl lactate. Solketal can be used in plasticizers and paints, as a carrier for some injectable veterinary treatments and insecticides [117], and as a solvent in ink formulations, cooling agents and cleaning products [118]. Furthermore, the acetate esters of solketal and glycerol formal have recently been reported as a potential biofuel additive [119]. However, they are characterized as medium to high dipolarity, which suggests that they could be a replacement in certain processes that require a polar aprotic solvent, although there have been no reports of their use so far in the literature.

Many glycerol ethers have also gained attention as potential low VOC solvents [120]. 1,2,3-Trimethoxypropane has similar solubility properties to ethers, esters and DCM due to its lack of hydrogen-bonding capability compared to glycerol. New cleaner synthetic methods have recently been established for its synthesis and it has been tested in the Heck, Suzuki and Sonogashira C–C coupling reactions, as well as Grignard and Barbier organometallic reactions [121]. 1,2,3-Trimethoxypropane has also been reported as the reaction medium for the reduction of nitro, nitrile, ester and acid functional groups by reaction with 1,1,3,3-tetramethyldisiloxane (TMDS) and using copper catalysts, with good yields obtained [122]. Many polymers are also soluble in 1,2,3-trimethoxypropane, such as cellulose acetate, poly(vinyl chloride), polyester, polyethylene and polycaprolactone [121]. It is not readily biodegradable but has low toxicity and a high flash point [122].

All of the above solvents have found limited uses as of yet but all have interesting solubility parameters, can be sourced from waste glycerol, and are higher-boiling alternatives to many problematic solvents used today. Of course, the low volatility of many bio-based solvents is hard to reconcile with contemporary distillation approaches to solvent removal, so a shift towards glycerol-derived solvents in processing is dependent on the development of improved solvent recovery technologies.

3.4   Terpenes and Derivatives

Terpenes are a diverse type of hydrocarbon that are found as major components in essential oils and resins of a diverse range of biomass. Importantly, they can be sourced from plant waste such as citrus peels and pine resins.

Citrus fruits are extensively produced worldwide, particularly in Brazil, the United States, China, Japan, Spain and South Africa. Over 88 million tonnes of the citrus fruits orange, lime, tangerine, lemon and grapefruit are produced annually [123] and nearly half of these citrus fruits are utilized for fruit juice production [124]. The citrus peels obtained during the processing of citrus fruits account for half of the raw fruit and are discarded as waste [125]. This has been estimated to amount to about 15.6 million tonnes of citrus waste every year [4].

Turpentine is already well established as a solvent for paint and varnish thinning, and is composed of a mix of terpenes. It is sourced from pine trees and can be considered a waste as it is a by-product of the forestry industry and does not compete with food supplies. There are different kinds of turpentine depending on the means of production: sulfate and sulfite turpentine are by-products from the Kraft and sulfite processes, respectively; turpentine oil is distilled from the crude sap that can be tapped from living pine trees; steam-distilled turpentine is produced by the steam distillation of wood chips, a by-product from lumbering; and destructively distilled wood turpentine is produced by the dry distillation of waste pinewood such as tree stumps [126].

The two most commonly used terpenes in the solvent sector are D-limonene and the pinene isomers, which form the main component of turpentine. Terpenes can be further modified to produce more stable derivatives, such as in the case of *p*-cymene from D-limonene. This section will focus on the terpenes: D-limonene, *p*-cymene, *p*-menthane, eucalyptol and the pinene isomers, which for simplicity will be collectively referred to as pinene (Figure 3.10).

D-Limonene ((*R*)-(+)-limonene) is the main component of citrus oils [127]. Taking orange peel waste as an example, the essential oil comprises around 4% of its dry mass, with D-limonene the primary component (about 95% by weight) [128]. D-Limonene is a biodegradable, high-boiling monocyclic monoterpene of low polarity with a distinctive odour. It has been classified as suitable for the production of fragrances and flavours by the FDA, in which it is a well-established ingredient. The full potential of citrus waste as a bountiful resource is only beginning to be realized, but a potentially sustainable supply of D-limonene should be considered as an extremely valuable source of solvents and other bio-based chemicals [4].

Much research has been dedicated to the production of D-limonene from citrus waste. It is extracted from citrus peels by distillation using cold pressing, or Clevenger apparatus [129]. A significant advantage of these methods is that the food industry potential of the remaining waste is unaffected (e.g. for subsequent pectin extraction). D-Limonene can also be extracted from orange peels by either steam distillation or microwave-assisted steam diffusion [82, 130]. Balu *et al.* [131]

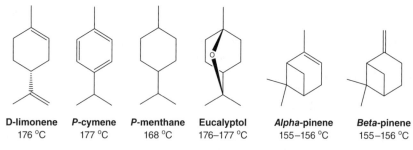

D-limonene	**P-cymene**	**P-menthane**	**Eucalyptol**	*Alpha*-**pinene**	*Beta*-**pinene**
176 °C	177 °C	168 °C	176–177 °C	155–156 °C	155–156 °C

Figure 3.10   Various terpenes and derivatives with solvent applications and their corresponding boiling points.

demonstrated the value of microwave technology when producing D-limonene in this way. Pourbafrani *et al.* [128] reported the production of D-limonene from treatment of orange peels using H_2SO_4 at 150°C in a cost-effective biorefinery model process. Mira *et al.* [132] proposed a supercritical CO_2 extraction method for the production of D-limonene, obtaining a yield of 5% based on the dry weight of the orange peels. Carlson *et al.* [133] also developed a method of extraction of D-limonene from citrus essential oil using supercritical CO_2 and membranes. Kulkarni *et al.* [129] reported the development of an integrated process that uses poly(propylene glycol) 240 (PPG) with subsequent organophilic pervaporation to selectively release D-limonene. Commercial processes are operational in Brazil and Florida for the production of D-limonene, co-produced as a side product during the large-scale production of citrus juices [134].

Chemat *et al.* [135] reported the use of D-limonene as a bio-based extraction solvent to replace *n*-hexane in the extraction of oils from olive seed. The extraction was conducted in a microwave-integrated Soxhlet extractor [135], following which D-limonene was removed from the extracted oil by aqueous azeotropic distillation assisted by microwave Clevenger apparatus. Although D-limonene has a higher boiling point (176°C) than that of *n*-hexane (69°C), increasing energy use during the solvent removal, employing azeotropic distillation decreases energy consumption dramatically, with the azeotrope distilling below 100°C. D-Limonene was also utilized for the extraction of simvastatin and lovastatin as well as their hydroxyacid metabolites from human plasma samples [136]. Another example of D-limonene assisted extraction is that of rice bran oil, with comparable oil quality to that achieved with *n*-hexane as well as the advantage of lower regulatory concerns and the production of a food-safe product [137]. Chemat-Djenni *et al.* [138] demonstrated the suitability of bio-derived D-limonene steam-distilled from orange peel waste for the extraction of lycopene. A highly hydrophobic nutraceutical was successfully obtained utilizing a food-safe solvent with similar results to those seen with DCM, which is the recommended solvent for extraction of lycopene. Dejoye Tanzi *et al.* [139] reported that they used D-limonene in place of *n*-hexane in the extraction of oils from microalgae. Shin and Chase [140] demonstrated the use of D-limonene to dissolve polystyrene for electrospun recycling in place of more traditional solvents such as THF, DMF and dimethylacetamide (DMAc).

Although D-limonene exhibits many advantages over *n*-hexane, for example, such as being bio-based and biodegradable with low human toxicity, there are still some obvious problems arising from its application as a solvent. The availability of the citrus waste feedstock is closely linked to crop production, which in some locations is seasonal, but more generally can be affected by drought and fluctuating demand. The high boiling point of D-limonene (176°C) makes separation more energy-intensive than the conventional lower-boiling solvents such as DCM and *n*-hexane, and its price is higher than many petrochemical solvents. The reactive alkene of D-limonene prevents it from being used under some reaction conditions, and although not toxic to humans it is toxic to aquatic life [141].

Nevertheless, owing to the position of D-limonene as a low-polarity renewable solvent with the ability to replace unsustainable petroleum-based hydrocarbons, it is still considered to be a key bio-based solvent and is used widely. The application of D-limonene across many areas of the chemical industry (cleaning, degreasing, synthetic chemistry) is indicative of this [142, 143]. In the consumer sector, the plant origin of D-limonene enhances the perception of the product.

The oxidation of D-limonene to *p*-cymene has attracted attention (Figure 3.11), originally as a means of producing a precursor for terephthalic acid [144, 145], but now increasingly for the synthesis of a solvent, fragrance, flavouring and renewable *p*-cresol precursor [146]. Solid metal oxide catalysts can be utilized in microwave conditions for this process [147] and, unlike D-limonene, *p*-cymene is aromatic and hence more stable. It is an analogue of toluene but with a higher boiling point, and as such *p*-cymene has similarly been used in esterifications and amidations [148]. Another appealing aspect of the production of *p*-cymene is that renewable hydrogen is also produced, which could be collected for use in other chemical processes [147].

An additional solvent in this class of D-limonene derivatives is the aliphatic hydrocarbon *p*-menthane, the fully unsaturated analogue of D-limonene. It too is a low-polarity hydrocarbon and offers another option as an alternative to petroleum-derived hydrocarbons such as pentane and cyclohexane. At present, interest in *p*-cymene has overshadowed that in *p*-menthane, even though the palladium-catalysed transformation of D-limonene is actually a disproportionation yielding *p*-cymene and *p*-methane (Figure 3.12) [149].

Despite the present-day heightened interest in bio-based solvents, the use of pinene as a solvent has not extended far beyond traditional turpentine applications. However, owing to its very similar physical and solubility properties compared to D-limonene, it could be used as a limonene alternative in many of the applications mentioned above.

Eucalyptol is an interesting molecule structurally. Unlike the other terpenes discussed in this section, it is not unsaturated and so is not prone to polymerization. It is a bicyclic ether which exhibits higher basicity [150], lower polarity [151] and hence lower water miscibility compared to other cyclic ethers such as THF

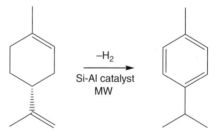

Figure 3.11 *p*-Cymene production from D-limonene using a silica–alumina catalyst and microwave irradiation.

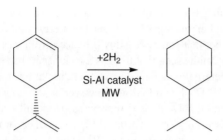

Figure 3.12   The synthesis of *p*-menthane from D-limonene.

Table 3.2   *Kamlet–Taft solvatochromic descriptors as well as Hansen solubility parameters of eucalyptol, 2-MeTHF and THF*

Solvent	Eucalyptol	2-MeTHF	THF
Boiling point (°C)	176–177	80	66
Acidity (α)	0	0	0
Basicity (β)	0.61	0.58	0.55
Polarity (π)	0.36	0.53	0.55
Hansen (δ_H)	3.4	4.3	8.0
Hansen (δ_P)	4.6	5.0	5.7
Hansen (δ_D)	16.7	16.9	16.8

or 2-MeTHF (Table 3.2). The combination of these properties could make it a very useful solvent for organometallic chemistry, where the absence of water is often essential. With a boiling point range of 176–177°C, it can facilitate higher reaction temperatures than typical ether solvents at the expense of facile removal. It is the major component of the essential oil from *Eucalyptus globulus* leaves, although because of a potentially difficult purification, a more likely source would be from the oxidative cyclization of D-limonene via α-terpineol. Eucalyptol is an alternative precursor for *p*-cymene synthesis [152]. At present, eucalyptol is too expensive for use as a general-purpose solvent, but its interesting solvent properties could encourage research into its synthesis from D-limonene.

3.5   Conclusion

It is no small task to find replacements for traditional petroleum-derived solvents from biomass waste. The large scale of demand, the wide range of applications and an array of necessary solvent properties must all be considered. However, progress is being made in qualifying and quantifying available waste sources, identifying potential routes to solvents from these wastes and their subsequent manufacture, and testing and characterizing new solvents for their suitability in different reactions and conditions. A solvent polarity map (arranged according to the dipolarity and hydrogen-bonding Hansen solubility parameters) shows how far the state of the art has progressed in this regard, with most property combinations represented by a bio-based solvent that can be feasibly produced from waste sources.

In Figure 3.13, it can be seen that there is a lack of bio-based solvents with the same polarity characteristics as the conventional dipolar aprotic class of solvents: namely DMAc, NMP, DMF and DMSO. However, it is encouraging that research in the field of bio-based solvents is still growing. Hopefully, the relative absence of viable substitutes in this area can be addressed sooner rather than later, with more bio-based solvents complementing GVL and cyrene. Another area in which there is little bio-representation is low-boiling, low-polarity solvents. Low-polarity bio-solvents such as limonene and *p*-cymene are high-boiling and, while they make valuable contributions to filling the solvent space, some low-boiling alternatives must also be found to complement them.

In most cases, the final barrier to the large-scale production of many of the alternative green solvents presented in this discussion is an economic one. We must be able to produce these bio-based solvents from waste at a cost that is competitive with that of their petroleum-derived counterparts. Otherwise, assuming regulatory compliance, the only immediate incentive for industry to change comes from consumer pressure. With increased research efforts addressing the subject of solvent

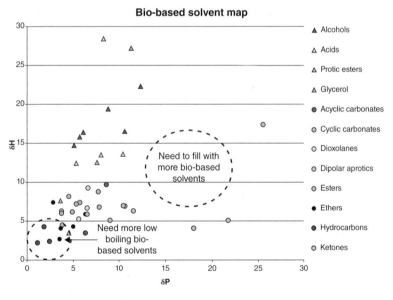

Figure 3.13 Map of current bio-based solvents using HSP hydrogen-bonding versus polarity scales. Current bio-based hydrocarbons (purple circles) are high-boiling. There is a need to develop lower-boiling bio-derived solvents as alternatives. There is also an empty space where dipolar aprotic solvents such as NMP, DMF and DMSO would be located, which must be filled with bio-based solvents. (A colour version of this figure appears in the plate section.)

production from plausible (waste) feedstocks, and not just exploring applications for solvents that may well remain academic curiosities, the scientific and engineering basis for a bio-based solvent industry is becoming a more realistic goal. A further push is provided by the present-day political unrest in the Middle East and the dwindling resources of easily exploitable petroleum, at least by conventional methods. At the time of writing, the low cost of shale gas and a market surplus of crude oil might appear to have thwarted hopes of a bio-based economy. Nevertheless, bio-based products are successful today regardless because consumers appreciate the environmental benefit they provide. The size and value of the bio-based market still is projected to increase [153, 154], and meanwhile short-term oil price fluctuations cannot continue to mask the finite nature of the petrochemical industry.

References

1. European Solvents Industry Group (2015) *Solvents facts and figures*. http://www.esig.org/en/about-solvents/what-are-solvents/facts-and-figures-the-european-solvents-industry-in-brief (accessed 28 August 2015).
2. Clark, J.H., Farmer, T.J., Hunt, A.J. and Sherwood, J. (2015) Opportunities for bio-based solvents created as petrochemical and fuel products transition towards renewable resources. *Int. J. Mol. Sci.*, **16** (8), 17101–17159.
3. Fritsche, U.R., Sims, R.E.H. and Monti, A. (2010) Direct and indirect land-use competition issues for energy crops and their sustainable production – an overview. *Biofuels Bioprod. Biorefining*, **4** (6), 692–704.
4. Lin, C.S.K., Pfaltzgraff, L.A., Herrero-Davila, L., *et al.* (2013) Food waste as a valuable resource for the production of chemicals, materials and fuels. Current situation and global perspective. *Energy Environ. Sci.*, **6** (2), 426–464.
5. Bio Intelligence Service, European Commission (2010) *Preparatory study on food waste across EU* 27. http://ec.europa.eu/environment/eussd/pdf/bio_foodwaste_report.pdf (accessed 28 August 2015).
6. UN News Service Section (2013) *UN News – UN report: One-third of world's food wasted annually, at great economic, environmental cost*. http://www.un.org/apps/news/story.asp?NewsID=45816#.VeDC0fZViko (accessed 28 August 2015).
7. Clark, J.H., Pfaltzgraff, L.A., Budarin, V.L., *et al.* (2013) From waste to wealth using green chemistry. *Pure Appl. Chem.*, **85** (8), 1625–1631.
8. Martin-Luengo, M.A., Yates, M., Rojo, E.S., *et al.* (2010) Sustainable *p*-cymene and hydrogen from limonene. *Appl. Catal. A: Gen.*, **387** (1–2), 141–146.
9. Azadi, P., Carrasquillo-Flores, R., Pagán-Torres, Y.J., *et al.* (2012) Catalytic conversion of biomass using solvents derived from lignin. *Green Chem.*, **14** (6), 1573–1576.
10. Hunt, A.J., Sin, E.H.K., Marriott, R. and Clark, J.H. (2010) Generation, capture, and utilization of industrial carbon dioxide. *ChemSusChem*, **3** (3), 306–322.
11. Leitner, W. (2002) Supercritical carbon dioxide as a green reaction medium for catalysis. *Acc. Chem. Res.*, **35** (9), 746–756.
12. North, M., Pasquale, R. and Young, C. (2010) Synthesis of cyclic carbonates from epoxides and CO_2. *Green Chem.*, **12** (9), 1514–1539.

13. Gu, Y. and Jérôme, F. (2010) Glycerol as a sustainable solvent for green chemistry. *Green Chem.*, **12** (7), 1127–1138.
14. Hu, J., Du, Z., Tang, Z. and Min, E. (2004) Study on the solvent power of a new green solvent: biodiesel. *Ind. Eng. Chem. Res.*, **43** (24), 7928–7931.
15. Moity, L., Benazzouz, A., Molinier, V., *et al.* (2015) Glycerol acetals and ketals as bio-based solvents: positioning in Hansen and COSMO-RS spaces, volatility and stability towards hydrolysis and autoxidation. *Green Chem.*, **17** (3), 1779–92.
16. Carrier, M., Loppinet-Serani, A., Denux, D., *et al.* (2011) Thermogravimetric analysis as a new method to determine the lignocellulosic composition of biomass. *Biomass Bioenergy*, **35** (1), 298–307.
17. Cherubini, F., Jungmeier, G., Wellisch, M., *et al.* (2009) Toward a common classification approach for biorefinery systems. *Biofuels Bioprod. Biorefining*, **3** (5), 534–546.
18. Yang, R., Yu, X., Zhang, Y., *et al.* (2008) A new method of low-temperature methanol synthesis on Cu/ZnO/Al$_2$O$_3$ catalysts from CO/CO$_2$/H$_2$. *Fuel*, **87** (4–5), 443–450.
19. Yang, B. and Wyman, C.E. (2008) Pretreatment: the key to unlocking low-cost cellulosic ethanol. *Biofuels Bioprod. Biorefining*, **2** (1), 26–40.
20. Hayes, D.J., Fitzpatrick, S., Hayes, M.H.B. and Ross, J.R.H. (2005) The Biofine process – production of levulinic acid, furfural, and formic acid from lignocellulosic feedstocks. In *Biorefineries – Industrial Processes and Products: Status Quo and Future Directions* (eds B. Kamm, P.R. Gruber and M. Kamm), Wiley-VCH, Weinheim, pp. 139–164.
21. Sherwood, J., Bruyn, M.D., Constantinou, A., *et al.* (2014) Dihydrolevoglucosenone (cyrene) as a bio-based alternative for dipolar aprotic solvents. *Chem. Commun.*, **50** (68), 9650–9652.
22. Zheng, Y., Zhao, J., Xu, F. and Li, Y. (2014) Pretreatment of lignocellulosic biomass for enhanced biogas production. *Prog. Energy Combust. Sci.*, **42** (June), 35–53.
23. Brandt, A., Gräsvik, J., Hallett, J.P. and Welton, T. (2013) Deconstruction of lignocellulosic biomass with ionic liquids. *Green Chem.*, **15** (3), 550–583.
24. Sitepu, I.R., Shi, S., Simmons, B.A., *et al.* (2014) Yeast tolerance to the ionic liquid 1-ethyl-3-methylimidazolium acetate. *FEMS Yeast Res.*, **14** (8), 1286–1294.
25. Santos, A.G., Ribeiro, B.D., Alviano, D.S. and Coelho, M.A.Z. (2014) Toxicity of ionic liquids toward microorganisms interesting to the food industry. *RSC Adv.*, **4** (70), 37157–37163.
26. Thuy Pham, T.P., Cho, C.-W. and Yun, Y.-S. (2010) Environmental fate and toxicity of ionic liquids: a review. *Water Res.*, **44** (2), 352–372.
27. Sun, Y. and Cheng, J. (2002) Hydrolysis of lignocellulosic materials for ethanol production: a review. *Bioresour. Technol.*, **83** (1), 1–11.
28. Sánchez, C. (2009) Lignocellulosic residues: biodegradation and bioconversion by fungi. *Biotechnol. Adv.*, **27** (2), 185–194.
29. Aycock, D.F. (2006) Solvent applications of 2-methyltetrahydrofuran in organometallic and biphasic reactions. *Org. Process Res. Dev.*, **11** (1), 156–159.
30. Parshetti, G.K., Suryadharma, M.S., Thuy Pham, T.P., *et al.* (2015) Heterogeneous catalyst-assisted thermochemical conversion of food waste biomass into 5-hydroxymethylfurfural. *Bioresour. Technol.*, **178**, 19–27.
31. Upare, P.P., Lee, J.-M., Hwang, Y.K., *et al.* (2011) Direct hydrocyclization of biomass-derived levulinic acid to 2-methyltetrahydrofuran over nanocomposite copper/silica catalysts. *ChemSusChem*, **4** (12), 1749–1752.

32. Antonucci, V., Coleman, J., Ferry, J.B., *et al.* (2011) Toxicological assessment of 2-methyl tetrahydrofuran and cyclopentyl methyl ether in support of their use in pharmaceutical chemical process development. *Org. Process Res. Dev.*, **15** (4), 939–941.

33. Penn Speciality Chemicals (2005) *Methyltetrahydrofuran.* http://www.pennakem.com/pdfs/methf5.pdf (accessed 29 September 2015).

34. Kadam, A., Nguyen, M., Kopach, M., *et al.* (2013) Comparative performance evaluation and systematic screening of solvents in a range of Grignard reactions. *Green Chem.*, **15** (7), 1880–1888.

35. Kawakami, J., Nakamoto, K., Nuwa, S., *et al.* (2010) Process for producing fused imidazole compound, Reformatsky reagent in stable form, and process for producing the same. *Patent US* **7662974** B2. http://www.google.com/patents/US7662974 (accessed 24 August 2015).

36. Rathman, T.L. and Schwindeman, J.A. (2014) Preparation, properties, and safe handling of commercial organolithiums: alkyllithiums, lithium *sec*-organoamides, and lithium alkoxides. *Org. Process Res. Dev.*, **18** (10), 1192–1210.

37. Krishnan, S. and Schreiber, S.L. (2004) Syntheses of stereochemically diverse nine-membered ring-containing biaryls. *Org. Lett.*, **6** (22), 4021–4024.

38. MacMillan, D.S., Murray, J., Sneddon, H.F., *et al.* (2013) Evaluation of alternative solvents in common amide coupling reactions: replacement of dichloromethane and *N,N*-dimethyl formamide. *Green Chem.*, **15** (3), 596.

39. Milton, E.J. and Clarke, M.L. (2010) Palladium-catalysed Grignard cross-coupling using highly concentrated Grignards in methyl-tetrahydrofuran. *Green Chem.*, **12** (3), 381–383.

40. Smoleń, M., Kędziorek, M. and Grela, K. (2014) 2-Methyltetrahydrofuran: sustainable solvent for ruthenium-catalyzed olefin metathesis. *Catal. Commun.*, **44**, 80–84.

41. Zhong, W., Wu, Y. and Zhang, X. (2009) Efficient chemoselective addition of Grignard reagents to carbonyl compounds in 2-methyltetrahydrofuran. *J. Chem. Res.*, **2009** (6), 370–373.

42. Robert, T., Velder, J. and Schmalz, H.-G. (2008) Enantioselective Cu-catalyzed 1,4-addition of Grignard reagents to cyclohexenone using taddol-derived phosphine–phosphite ligands and 2-methyl-THF as a solvent. *Angew. Chem., Int. Edn*, **47** (40), 7718–7721.

43. Kawabata, S., Tokura, H., Chiyojima, H., *et al.* (2012) Asymmetric hydrosilane reduction of ketones catalyzed by an iridium complex bearing a hydroxyamide-functionalized NHC ligand. *Adv. Synth. Catal.*, **354** (5), 807–812.

44. vom Stein, T., Grande, P.M., Leitner, W. and Domínguez de María, P. (2011) Iron-catalyzed furfural production in biobased biphasic systems: from pure sugars to direct use of crude xylose effluents as feedstock. *ChemSusChem*, **4** (11), 1592–1594.

45. Chen, Z.-G., Zhang, D.-N., Cao, L. and Han, Y.-B. (2013) Highly efficient and regioselective acylation of pharmacologically interesting cordycepin catalyzed by lipase in the eco-friendly solvent 2-methyltetrahydrofuran. *Bioresour. Technol.*, **133**, 82–86.

46. Duan, Z.-Q. and Hu, F. (2013) Efficient synthesis of phosphatidylserine in 2-methyltetra hydrofuran. *J. Biotechnol.*, **163** (1), 45–49.

47. Shanmuganathan, S., Natalia, D., van den Wittenboer, A., *et al.* (2010) Enzyme-catalyzed C–C bond formation using 2-methyltetrahydrofuran (2-MTHF) as (co)solvent: efficient and bio-based alternative to DMSO and MTBE. *Green Chem.*, **12** (12), 2240–2245.

48. Grochowski, M.R., Yang, W. and Sen, A. (2012) Mechanistic study of a one-step catalytic conversion of fructose to 2,5-dimethyltetrahydrofuran. *Chemistry, Eur. J.*, **18** (39), 12363–12371.

49. Fábos, V., Koczó, G., Mehdi, H., *et al.* (2009) Bio-oxygenates and the peroxide number: a safety issue alert. *Energy Environ. Sci.*, **2** (7), 767.

50. Court, G.R., Lawrence, C.H., Raverty, W.D. and Duncan, A.J. (2011) Method for converting lignocellulosic materials into useful chemicals. Patent WO 2011000030 A1. http:// www.google.com/patents/WO2011000030A1 (accessed 30 August 2015).

51. Horváth, I.T., Mehdi, H., Fábos, V., *et al.* (2008) γ-Valerolactone – a sustainable liquid for energy and carbon-based chemicals. *Green Chem.*, **10** (2), 238.

52. Antal, Jr.,, M.J., Mok, W.S.L. and Richards, G.N. (1990) Mechanism of formation of 5-(hydroxymethyl)-2-furaldehyde from *d*-fructose and sucrose. *Carbohydr. Res.*, **199** (1), 91–109.

53. Alonso, D.M., Wettstein, S.G. and Dumesic, J.A. (2013) Gamma-valerolactone, a sustainable platform molecule derived from lignocellulosic biomass. *Green Chem.*, **15** (3), 584–595.

54. Qi, L. and Horváth, I.T. (2012) Catalytic conversion of fructose to γ-valerolactone in γ-valerolactone. *ACS Catal.*, **2** (11), 2247–2249.

55. Alonso, D.M., Wettstein, S.G., Mellmer, M.A., *et al.* (2012) Integrated conversion of hemicellulose and cellulose from lignocellulosic biomass. *Energy Environ. Sci.*, **6** (1), 76–80.

56. Strappaveccia, G., Ismalaj, E., Petrucci, C., *et al.* (2014) A biomass-derived safe medium to replace toxic dipolar solvents and access cleaner Heck coupling reactions. *Green Chem.*, **17** (1), 365–372.

57. Strappaveccia, G., Luciani, L., Bartollini, E., *et al.* (2015) γ-Valerolactone as an alternative biomass-derived medium for the Sonogashira reaction. *Green Chem.*, **17** (2), 1071–1076.

58. Ismalaj, E., Strappaveccia, G., Ballerini, E., *et al.* (2014) γ-Valerolactone as a renewable dipolar aprotic solvent deriving from biomass degradation for the Hiyama reaction. *ACS Sustain. Chem. Eng.*, **2** (10), 2461–2464.

59. Qi, L., Mui, Y.F., Lo, S.W., *et al.* (2014) Catalytic conversion of fructose, glucose, and sucrose to 5-(hydroxymethyl)furfural and levulinic and formic acids in γ-valerolactone as a green solvent. *ACS Catal.*, **4** (5), 1470–1477.

60. Gallo, J.M.R., Alonso, D.M., Mellmer, M.A. and Dumesic, J.A. (2012) Production and upgrading of 5-hydroxymethylfurfural using heterogeneous catalysts and biomass-derived solvents. *Green Chem.*, **15** (1), 85–90.

61. GF Biochemicals (2015) *Derivatives: Levulinic acid esters (LA-esters)*. http://www .gfbiochemicals.com/products/#derivatives (accessed 21 September 2015).

62. Tundo, P., Aricò, F., Gauthier, G., *et al.* (2010) Green synthesis of dimethyl isosorbide. *ChemSusChem*, **3** (5), 566–570.

63. Lavergne, A., Zhu, Y., Pizzino, A., *et al.* (2011) Synthesis and foaming properties of new anionic surfactants based on a renewable building block: sodium dodecyl isosorbide sulfates. *J. Colloid Interface Sci.*, **360** (2), 645–653.

64. Breffa, C., Beckedahl, B., Dierker, M., *et al.* (2011) Use of isosorbide ethers in detergents and cleaners. *Patent US* **8008246** B2. http://www.google.com/patents/US8008246 (accessed 24 August 2015).

65. Mouret, A., Leclercq, L., Mühlbauer, A. and Nardello-Rataj, V. (2013) Eco-friendly solvents and amphiphilic catalytic polyoxometalate nanoparticles: a winning combination for olefin epoxidation. *Green Chem.*, **16** (1), 269–278.

66. Tan, J.-N., Ahmar, M. and Queneau, Y. (2013) HMF derivatives as platform molecules: aqueous Baylis–Hillman reaction of glucosyloxymethyl-furfural towards new biobased acrylates. *RSC Adv.*, **3** (39), 17649–17653.

67. Zhang, K., Sawaya, M.R., Eisenberg, D.S. and Liao, J.C. (2008) Expanding metabolism for biosynthesis of nonnatural alcohols. *Proc. Natl. Acad. Sci.*, **105** (52), 20653–20658.
68. Atsumi, S., Hanai, T. and Liao, J.C. (2008) Non-fermentative pathways for synthesis of branched-chain higher alcohols as biofuels. *Nature*, **451** (7174), 86–89.
69. Zhu, T., Xie, X., Li, Z., *et al.* (2014) Enhancing photosynthetic production of ethylene in genetically engineered *Synechocystis* sp. PCC 6803. *Green Chem.*, **17** (1), 421–434.
70. van Leeuwen, B.N.M., van der Wulp, A.M., Duijnstee, I., *et al.* (2012) Fermentative production of isobutene. *Appl. Microbiol. Biotechnol.*, **93** (4), 1377–1387.
71. Koutinas, A.A., Vlysidis, A., Pleissner, D., *et al.* (2014) Valorization of industrial waste and by-product streams via fermentation for the production of chemicals and biopolymers. *Chem. Soc. Rev.*, **43** (8), 2587–2627.
72. Blombach, B. and Eikmanns, B.J. (2011) Current knowledge on isobutanol production with *Escherichia coli, Bacillus subtilis* and *Corynebacterium glutamicum. Bioengineered*, **2** (6), 346–350.
73. Taylor, T.J., Taylor, J.D., Peters, M.W. and Henton, D.E. (2012) Variations on Prins-like chemistry to produce 2,5-dimethylhexadiene from isobutanol. Patent US 20120271082 A1. https://www.google.com/patents/US20120271082 (accessed 22 March 2015).
74. Bockrath, R. (2014) Improved fermentation method. Patent WO 2014086780 A2. http://www.google.com/patents/WO2014086780A2 (accessed 3 August 2015).
75. Global Bioenergies (2015) *The isobutene process successfully uses xylose, the 'wood sugar'.* http://www.global-bioenergies.com/the-isobutene-process-successfully-uses-xylose-the-wood-sugar/?lang=en (accessed 30 August 2015).
76. Zhang, C.M., Adesina, A.A. and Wainwright, M.S. (2003) Isobutene hydration over Amberlyst-15 in a slurry reactor. *Chem. Eng. Process., Process Intensif.*, **42** (12), 985–991.
77. Salavati-Niasari, M., Khosousi, T. and Hydarzadeh, S. (2005) Highly selective esterification of *tert*-butanol by acetic acid anhydride over alumina-supported InCl$_3$, GaCl$_3$, FeCl$_3$, ZnCl$_2$, CuCl$_2$, NiCl$_2$, CoCl$_2$ and MnCl$_2$ catalysts. *J. Mol. Catal. Chem.*, **235** (1–2), 150–153.
78. Bergman, C. and Millett, J. (2015) *After extensive analysis, EPA removes chemicals from lists of regulated pollutants.* http://yosemite.epa.gov/opa/admpress.nsf/a21708abb48b5 a9785257359003f0231/50d1d8b63a857ce785256f500065d12e!OpenDocument (accessed 23 September 2015).
79. Henderson, R.K., Jiménez-González, C., Constable, D.J.C., *et al.* (2011) Expanding GSK's solvent selection guide – embedding sustainability into solvent selection starting at medicinal chemistry. *Green Chem.*, **13** (4), 854–862.
80. Zheng, P., Dong, J.-J., Sun, Z.-H., *et al.* (2009) Fermentative production of succinic acid from straw hydrolysate by *Actinobacillus succinogenes. Bioresour. Technol.*, **100** (8), 2425–2429.
81. Leung, C.C.J., Cheung, A.S.Y., Zhang, A.Y.-Z., *et al.* (2012) Utilisation of waste bread for fermentative succinic acid production. *Biochem. Eng. J.*, **65**, 10–15.
82. Li, Q., Siles, J.A. and Thompson, I.P. (2010) Succinic acid production from orange peel and wheat straw by batch fermentations of *Fibrobacter succinogenes* S85. *Appl. Microbiol. Biotechnol.*, **88** (3), 671–678.
83. Sun, Z., Li, M., Qi, Q., *et al.* (2014) Mixed food waste as renewable feedstock in succinic acid fermentation. *Appl. Biochem. Biotechnol.*, **174** (5), 1822–1833.
84. Corbion (2015) *Purac.* http://www.corbion.com/bioplastics/about-bioplastics/raw-material-sources (accessed 23 September 2015).

85. Corma, A., Iborra, S. and Velty, A. (2007) Chemical routes for the transformation of biomass into chemicals. *Chem. Rev.*, **107** (6), 2411–2502.

86. NatureWorks (2014) *DOE awards $2.5 million to NatureWorks to transform biogas into the lactic acid building block for Ingeo*. http://www.natureworksllc.com/news-and-events/press-releases/2014/10-30-14-doe-grant-to-natureworks-to-transform-biogas-into-lactic-acid-for-ingeo (accessed 23 September 2015).

87. Liang, S., McDonald, A.G. and Coats, E.R. (2014) Lactic acid production with undefined mixed culture fermentation of potato peel waste. *Waste Manag.*, **34** (11), 2022–2027.

88. Wu, Y., Ma, H., Zheng, M. and Wang, K. (2015) Lactic acid production from acidogenic fermentation of fruit and vegetable wastes. *Bioresour. Technol.*, **191**, 53–58.

89. Probst, M., Walde, J., Pümpel, T., *et al.* (2015) A closed loop for municipal organic solid waste by lactic acid fermentation. *Bioresour. Technol.*, **175**, 142–151.

90. Li, X., Chen, Y., Zhao, S., *et al.* (2015) Efficient production of optically pure *l*-lactic acid from food waste at ambient temperature by regulating key enzyme activity. *Water Res.*, **70**, 148–157.

91. Datta, R. and Henry, M. (2006) Lactic acid: recent advances in products, processes and technologies — a review. *J. Chem. Technol. Biotechnol.*, **81** (7), 1119–1129.

92. Komesu, A., Martinez, P.F.M., Lunelli, B.H., *et al.* (2015) Lactic acid purification by reactive distillation system using design of experiments. *Chem. Eng. Process., Process Intensif.*, **95**, 26–30.

93. Sikder, J., Chakraborty, S., Pal, P., *et al.* (2012) Purification of lactic acid from microfiltrate fermentation broth by cross-flow nanofiltration. *Biochem. Eng. J.*, **69**, 130–137.

94. Yang, J., Tan, J.-N. and Gu, Y. (2012) Lactic acid as an invaluable bio-based solvent for organic reactions. *Green Chem.*, **14** (12), 3304–3317.

95. Clark, J.H., Macquarrie, D.J. and Sherwood, J. (2013) The combined role of catalysis and solvent effects on the Biginelli reaction: improving efficiency and sustainability. *Chemistry, Eur. J.*, **19** (16), 5174–5182.

96. Pereira, C.S.M., Pinho, S.P., Silva, V.M.T.M. and Rodrigues, A.E. (2008) Thermodynamic equilibrium and reaction kinetics for the esterification of lactic acid with ethanol catalyzed by acid ion-exchange resin. *Ind. Eng. Chem. Res.*, **47** (5), 1453–1463.

97. Petrie, E. (2011) *The evolution of bio-based green solvents*. http://www.metalfinishing.com/view/19390/the-evolution-of-bio-based-green-solvents/ (accessed 23 September 2015).

98. Gu, Y. and Jérôme, F. (2013) Bio-based solvents: an emerging generation of fluids for the design of eco-efficient processes in catalysis and organic chemistry. *Chem. Soc. Rev.*, **42**, 9550–9570.

99. Aparicio, S. and Alcalde, R. (2009) The green solvent ethyl lactate: an experimental and theoretical characterization. *Green Chem.*, **11** (1), 65–78.

100. Dandia, A., Jain, A.K. and Laxkar, A.K. (2013) Ethyl lactate as a promising bio-based green solvent for the synthesis of spiro-oxindole derivatives via 1,3-dipolar cycloaddition reaction. *Tetrahedron Lett.*, **54** (30), 3929–3932.

101. Datta, R. and Tsai, S.-P. (1998) Esterification of fermentation-derived acids via pervaporation. Patent US 5723639 A. http://www.google.co.in/patents/US5723639 (accessed 9 September 2014).

102. Bennett, J.S. (2011) Green synthesis of aryl aldimines using ethyl lactate. Patent US 2011 0196174 A1. https://www.google.com/patents/US20110196174 (accessed 14 December 2013).

103. Nikles, S.M., Piao, M., Lane, A.M. and Nikles, D.E. (2001) Ethyl lactate: a green solvent for magnetic tape coating. *Green Chem.*, **3** (3), 109–113.

104. Paul, S. and Das, A.R. (2012) An efficient green protocol for the synthesis of coumarin fused highly decorated indenodihydropyridyl and dihydropyridyl derivatives. *Tetrahedron Lett.*, **53** (17), 2206–2210.

105. Wan, J.-P., Wang, C., Zhou, R. and Liu, Y. (2012) Sustainable H$_2$O/ethyl lactate system for ligand-free Suzuki–Miyaura reaction. *RSC Adv.*, **2** (23), 8789–8792.

106. Xu, Z., Jiang, Y., Zou, S. and Liu, Y. (2014) Bio-based solvent mediated synthesis of dihydropyrimidinthiones via Biginelli reaction. *Phosphorus Sulfur Silicon Relat. Elem.*, **189** (6), 791–795.

107. Liu, Y., Wang, H., Wang, C., *et al.* (2013) Bio-based green solvent mediated disulfide synthesis via thiol couplings free of catalyst and additive. *RSC Adv.*, **3** (44), 21369–21372.

108. Salerno, A. and Domingo, C. (2014) A novel bio-safe phase separation process for preparing open-pore biodegradable polycaprolactone microparticles. *Mater. Sci. Eng. C*, **42**, 102–110.

109. Stefanidis, S., Kalogiannis, K., Iliopoulou, E.F., *et al.* (2013) Mesopore-modified mordenites as catalysts for catalytic pyrolysis of biomass and cracking of vacuum gasoil processes. *Green Chem.*, **15** (6), 1647–1658.

110. Mondal, D., Chaudhary, J.P., Sharma, M. and Prasad, K. (2014) Simultaneous dehydration of biomass-derived sugars to 5-hydroxymethyl furfural (HMF) and reduction of graphene oxide in ethyl lactate: one pot dual chemistry. *RSC Adv.*, **4** (56), 29834–29839.

111. Ciriminna, R., Pina, C.D., Rossi, M. and Pagliaro, M. (2014) Understanding the glycerol market. *Eur. J. Lipid Sci. Technol.*, **116** (10), 1432–1439.

112. Pukale, D.D., Maddikeri, G.L., Gogate, P.R., *et al.* (2015) Ultrasound assisted transesterification of waste cooking oil using heterogeneous solid catalyst. *Ultrason. Sonochem.*, **22**, 278–286.

113. Gonzalez-Arellano, C., Parra-Rodriguez, L. and Luque, R. (2014) Mesoporous Zr–SBA-16 catalysts for glycerol valorization processes: towards biorenewable formulations. *Catal. Sci. Technol.*, **4** (8), 2287–2292.

114. Schnabel, G. (2013) Use of glycerol derivatives as solvent in agrochemical compositions. *Patent WO* **2013153030** A1. http://www.google.com/patents/WO2013153030A1 (accessed 26 August 2015).

115. Radics, U., Niclas, H.-J., Schuster, C. and Geisler, M. (2013) Solvent systems for *N*-((3(5)-methyl-1*H*-pyrazol-1-yl)methyl)acetamide, and their use for the treatment of urea-based/ammonium-containing fertilizers. Patent DE 102013022031 B3. http://www.google.com/patents/DE102013022031B3 (accessed 26 August 2015).

116. Bayarri, F.N., Castells, B.J., Echeverria, B.B. and Estevez, C.C. (2007) Process for the preparation of loperamide. *Patent WO* **2008080601** A3. https://www.google.com/patents/WO2008080601A3 (accessed 26 August 2015).

117. Díaz-Álvarez, A.E., Francos, J., Lastra-Barreira, B., *et al.* (2011) Glycerol and derived solvents: new sustainable reaction media for organic synthesis. *Chem. Commun.*, **47** (22), 6208–6227.

118. García, J.I., García-Marín, H. and Pires, E. (2014) Glycerol based solvents: synthesis, properties and applications. *Green Chem.*, **16** (3), 1007–1033.

119. Dodson, J.R., Leite, T.dC.M., Pontes N.S., *et al.* (2014) Green acetylation of solketal and glycerol formal by heterogeneous acid catalysts to form a biodiesel fuel additive. *ChemSusChem*, **7** (9), 2728–2734.

120. García, J.I., García-Marín, H., Mayoral, J.A. and Pérez, P. (2010) Green solvents from glycerol. Synthesis and physico-chemical properties of alkyl glycerol ethers. *Green Chem.*, **12** (3), 426–434.

121. Sutter, M., Dayoub, W., Métay, E., *et al.* (2013) 1,2,3-Trimethoxypropane and glycerol ethers as bio-sourced solvents from glycerol. Synthesis by solvent-free phase-transfer catalysis and utilization as an alternative solvent in chemical transformations. *ChemCatChem*, **5** (10), 2893–2904.

122. Sutter, M., Pehlivan, L., Lafon, R., *et al.* (2013) 1,2,3-Trimethoxypropane, a glycerol-based solvent with low toxicity: new utilization for the reduction of nitrile, nitro, ester, and acid functional groups with TMDS and a metal catalyst. *Green Chem.*, **15** (11), 3020–3026.

123. Marín, F.R., Soler-Rivas, C., Benavente-García, O., *et al.* (2007) By-products from different citrus processes as a source of customized functional fibres. *Food Chem.*, **100** (2), 736–741.

124. Wilkins, M.R., Widmer, W.W., Grohmann, K. and Cameron, R.G. (2007) Hydrolysis of grapefruit peel waste with cellulase and pectinase enzymes. *Bioresour. Technol.*, **98** (8), 1596–1601.

125. Silez Lopez, J.Á., Quiang, L. and Thompson, I. (2010) Biorefinery of waste orange peel. *Crit. Rev. Biotechnol.*, **30** (1), 63–69.

126. Haneke, K. (2002) *Turpentine (turpentine oil, wood turpentine, sulfate turpentine, sulfite turpentine) [8006-64-2]*. US Department of Health and Human Services. https://ntp.niehs.nih.gov/ntp/htdocs/chem_background/exsumpdf/turpentine_508.pdf (accessed 27 August 2015).

127. Sun, J. (2007) D-Limonene: safety and clinical applications. *Alt. Med. Rev.*, **12** (3), 259–264.

128. Pourbafrani, M., Forgács, G., Horváth, I.S., *et al.* (2010) Production of biofuels, limonene and pectin from citrus wastes. *Bioresour. Technol.*, **101** (11), 4246–4250.

129. Kulkarni, P.S., Brazinha, C., Afonso, C.A.M. and Crespo, J.G. (2010) Selective extraction of natural products with benign solvents and recovery by organophilic pervaporation: fractionation of D-limonene from orange peels. *Green Chem.*, **12** (11), 1990–1994.

130. Farhat, A., Fabiano-Tixier, A.-S., Maataoui, M.E., *et al.* (2011) Microwave steam diffusion for extraction of essential oil from orange peel: kinetic data, extract's global yield and mechanism. *Food Chem.*, **125** (1), 255–261.

131. Balu, A.M., Budarin, V., Shuttleworth, P.S., *et al.* (2012) Valorisation of orange peel residues: waste to biochemicals and nanoporous materials. *ChemSusChem*, **5** (9), 1694–1697.

132. Mira, B., Blasco, M., Berna, A. and Subirats, S. (1999) Supercritical CO_2 extraction of essential oil from orange peel. Effect of operation conditions on the extract composition. *J. Supercrit. Fluids*, **14** (2), 95–104.

133. Carlson, L.H.C., Bolzan, A. and Machado, R.A.F. (2005) Separation of *d*-limonene from supercritical CO_2 by means of membranes. *J. Supercrit. Fluids*, **34** (2), 143–147.

134. Kimball, D.A. (1999) *Citrus Processing: A Complete Guide*, 2nd edn. Aspen Publishers, Gaithersburg, MD.

135. Virot, M., Tomao, V., Ginies, C., *et al.* (2008) Green procedure with a green solvent for fats and oils' determination: microwave-integrated Soxhlet using limonene followed by microwave Clevenger distillation. *J. Chromatogr. A*, **1196–1197**, 147–152.

136. Medvedovici, A., Udrescu, S. and David, V. (20103) Use of a green (bio)solvent – limonene – as extractant and immiscible diluent for large volume injection in the RPLC-tandem MS

assay of statins and related metabolites in human plasma. *Biomed. Chromatogr.*, **27** (1), 48–57.

137. Mamidipally, P.K. and Liu, S.X. (2004) First approach on rice bran oil extraction using limonene. *Eur. J. Lipid Sci. Technol.*, **106** (2), 122–125.

138. Chemat-Djenni, Z., Ferhat, M.A., Tomao, V. and Chemat, F. (2010) Carotenoid extraction from tomato using a green solvent resulting from orange processing waste. *J. Essent. Oil Bear. Plants*, **13** (2), 139–147.

139. Dejoye Tanzi, C., Abert Vian, M., Ginies, C., *et al.* (2012) Terpenes as green solvents for extraction of oil from microalgae. *Molecules (Basel)*, **17** (7), 8196–8205.

140. Shin, C. and Chase, G.G. (2005) Nanofibers from recycle waste expanded polystyrene using natural solvent. *Polym. Bull.*, **55** (3), 209–215.

141. Sigma Aldrich (2015). *Safety data sheet: (R)-(+)-Limonene.* MSDS-62118. http://www.sigmaaldrich.com/MSDS/MSDS/DisplayMSDSPage.do?country=GB&language=en&productNumber=62118&brand=SIAL&PageToGoToURL=http://www.sigmaaldrich.com/catalog/search?term=limonene&interface=All&N=0&mode=match%20partialmax&lang=en&region=GB&focus=product (accessed 23 September 2015).

142. Klier, J., Tucker, C.J. and Strandburg, G.M. (1997) High water content, low viscosity, oil continuous microemulsions and emulsions, and their use in cleaning applications. Patent US 5811383 A. http://www.google.com/patents/US5811383 (accessed 23 September 2015).

143. Zaki, N.N. and Troxler, R.E. (2012) Solvent compositions for removing petroleum residue from a substrate and methods of use thereof. *Patent US* **8951952** B2. http://www.google.com/patents/US8951952 (accessed 23 September 2015).

144. Colonna, M., Berti, C., Fiorini, M., *et al.* (2011) Synthesis and radiocarbon evidence of terephthalate polyesters completely prepared from renewable resources. *Green Chem.*, **13** (9), 2543–2548.

145. Berti, C., Binassi, E., Colonna, M., *et al.* (2008) Bio-based terephthalate polyesters. Patent US 20100168461 A1. http://www.google.com/patents/US20100168461 (accessed 26 August 2015).

146. Dávila, J.A., Rosenberg, M. and Cardona, C.A. (2015) Techno-economic and environmental assessment of *p*-cymene and pectin production from orange peel. *Waste Biomass Valorization*, **6** (2), 253–261.

147. Martín-Luengo, M.A., Yates, M., Martínez Domingo, M.J., *et al.* (2008) Synthesis of *p*-cymene from limonene, a renewable feedstock. *Appl. Catal. B: Environ.*, **81** (3–4), 218–224.

148. Clark, J.H., Macquarrie, D.J. and Sherwood, J. (2012) A quantitative comparison between conventional and bio-derived solvents from citrus waste in esterification and amidation kinetic studies. *Green Chem.*, **14** (1), 90–93.

149. Lesage, P., Candy, J.P., Hirigoyen, C., *et al.* (1996) Selective dehydrogenation of dipentene (*R*-(+)-limonene) into paracymene on silica supported palladium assisted by α-olefins as hydrogen acceptor. *J. Mol. Catal. Chem.*, **112** (3), 431–435.

150. Jessop, P.G., Jessop, D.A., Fu, D. and Phan, L. (2012) Solvatochromic parameters for solvents of interest in green chemistry. *Green Chem.*, **14** (5), 1245–1259.

151. Benazzouz, A., Moity, L., Pierlot, C., *et al.* (2013) Selection of a greener set of solvents evenly spread in the Hansen space by space-filling design. *Ind. Eng. Chem. Res.*, **52** (47), 16585–16597.

152. Leita, B.A., Warden, A.C., Burke, N., *et al.* (2010) Production of *p*-cymene and hydrogen from a bio-renewable feedstock – 1,8-cineole (eucalyptus oil). *Green Chem.*, **12** (1), 70–76.
153. de Besi, M. and McCormick, K. (2015) Towards a bioeconomy in Europe: national, regional and industrial strategies. *Sustainability*, **7** (8), 10461–10478.
154. Golden, J.S., Handfield, R.B., Daystar, J. and McConnell, T.E. (2015) *An Economic Impact Analysis of the U.S. Biobased Products Industry: A Report to the Congress of the United States of America*. United States Department of Agriculture. http://www.biopreferred.gov/BPResources/files/EconomicReport_6_12_2015.pdf (accessed 30 August 2015).

4

Deep Eutectic and Low-Melting Mixtures

Karine de Oliveira Vigier[1] and Joaquín García-Álvarez[2]

[1]*CNRS, Institut de Chimie des Milieux et Matériaux de Poitiers, Université de Poitiers, Poitiers, France*

[2]*CSIC, Laboratorio de Compuestos Organometálicos y Catálisis, Centro de Innovación en Química Avanzada, Universidad de Oviedo, Oviedo, Spain*

4.1 Introduction

With the end of the twentieth century, a new awareness arose to strive towards chemical processes that are aimed not only at productivity but also at ecological objectives [1]. In this sense, the so-called 'green chemistry' relies on 12 inherent principles, first postulated by Warner and Anastas, which define the subject and illustrate its objectives [2]. These include, among many others, the intentions of reducing wastes and increasing sustainability, energy efficiency and safety [3].

Besides unwanted by-products caused by stoichiometrically added reagents, waste also originates from solvents, which are necessary in the majority of reactions. Related to pharmaceutical syntheses, it has been estimated that up to 85% of the total involved mass is made up of solvents [4]. Furthermore, in these processes the recovery of the solvent generally ranges from 50% to 80% and thus causes enormous amounts of waste and consumes gigantic quantities of petroleum-based chemicals [5, 6].

Bio-Based Solvents, First Edition. Edited by François Jérôme and Rafael Luque.
© 2017 John Wiley & Sons Ltd. Published 2017 by John Wiley & Sons Ltd.

Despite the aforementioned low recovery rates, the depletion of crude petroleum and the fact that the vast majority of these solvents are volatile organic compounds (VOCs), with toxic or ecologically damaging properties, most chemical transformations are still performed in solutions of VOCs to: (i) achieve homogeneity of reagents and reactants; (ii) avoid the formation of undesired products by dilution, thus enhancing the reaction rate; (iii) ensure rapid and safe conversions; and (iv) efficiently control the heat flow of the reaction. (In this regard, a recent editorial [7] in *Organic Process Research & Development* discourages chemists from using solvents that are known to be toxic, are dangerous for large-scale preparations or are expensive to dispose of as waste.) To overcome the previously mentioned shortcomings of conventional VOCs, while still profiting from the positive effect of solvents [8], remarkable efforts have been focused on the replacement of traditional reaction media by green solvents.

To achieve these environment-friendly goals, there are basically two possibilities. Firstly, one can try to substitute highly toxic or hazardous solvents with conventional solvents that have better environmental, health and safety (EHS) properties [9]. But this clearly has its limits, since the solvents used can have specific effects on the yield and the selectivity of the reaction (e.g. organometallic compounds [10]). This leaves the second option, which is to try to find ecologically friendly solvents showing similar properties to the previous ones, like the best choice. Generally, green solvents are defined as such if they meet the following criteria: (i) availability; (ii) non-toxicity; (iii) biodegradability; (iv) recyclability; (v) non-flammability; and (vi) low price. For example, it is possible to eliminate the highly environmentally damaging chlorocarbon solvents in catalytic synthetic organic processes by using supercritical CO_2 (sc-CO_2) as reaction medium [11]. Furthermore, sc-CO_2 is relatively inert, easily removable and recyclable.

However, and due to the fact that reactions in supercritical media present several drawbacks [12] (like the requirement for special, expensive and high-pressure laboratory set-ups or their poor ability to solubilize many compounds), a different type of novel solvent, the ionic liquids (ILs), have received growing attention in the past decade [13, 14]. Put simply, ILs are liquids that are entirely composed of ions, which technically makes molten sodium chloride an ionic liquid. But, in general, the term 'ionic liquid' refers to a compound that is solely made up of ions and is liquid around or below 100°C [15]. Interestingly, compared to conventional organic compounds, ILs show a high thermal stability, non-flammability and practically no vapour pressure [16, 17]. This offers the possibility to substitute VOCs in today's synthesis with ILs and to reduce the environmental pollution caused by conventional organic solvents. Moreover, these non-volatile solvents allow an easy isolation of the final products through distillation. This idea of substituting conventional solvents with ILs is supported even further by the sheer number of possible combinations of ions to form ionic liquids. At the *International George Papatheodorou Symposium* in 1999, K. R. Seddon proposed that at least 10^6

binary and 10^8 ternary ILs are possible [15, 18]. Compared to the only 600 molecular solvents that are nowadays in use, ionic liquids open up great possibilities to tune solvents specifically to certain reactions to optimize yield, solubility and selectivity.

However, the universal application of ILs as green solvents has been reconsidered in current years. This is due to the fact that recent reports revealed a hazardous toxicity and poor biodegradability of most ILs (similar to those found on chlorinated solvents) [19]. On top of this, their synthesis is usually far from being environmentally friendly, as large amounts of salts and solvents are commonly required. In response to these drawbacks, the present focus lies in the development of new neoteric solvents that maximize the aforementioned green properties of traditional ILs, but avoid their negative ecological footprint. In this sense, recent pioneering work has recognized the potential of deep eutectic solvents [20] and low-melting mixtures [21] as biodegradable and biorenewable solvents, which offer new environmentally friendly and easily tunable reaction media for a large variety of chemical transformations.

Thus, this chapter is intended to cover part of the progress made during the last decade in the area of deep eutectic solvents (DESs) and low-melting mixtures (LMMs), providing a general overview of their application as green solvents in the fields of: (i) metal-catalysed organic reactions; and (ii) carbohydrate conversion and extraction processes. The importance of this subject is also increasing in other fields of chemistry, ranging from electrochemistry and organic synthesis to material sciences or metal extraction.

4.2   Deep Eutectic and Low-Melting Mixtures: Definition and Composition

The concept of deep eutectic solvents (DESs) was first introduced in 2003 by Abbott *et al.* [22] to describe the formation of a liquid eutectic mixture (melting point 12°C) by mixing two solid organic compounds: (i) choline chloride (ChCl, (2-hydroxyethyl)trimethylammonium chloride, melting point 133°C; Figure 4.1); and (ii) urea (melting point 302°C) in a 1 : 2 molar ratio. After this seminal work, in 2005 König and co-workers [23] reported the formation of the closely related low-melting mixtures (LMMs) based on bulk carbohydrates, sugar alcohols or citric acid (Figure 4.2) combined with different ureas and inorganic salts. In both cases, the eutectic solvent was formed as a result of the hydrogen-bonding interactions between the individual components of the mixture. Thus, DESs and LMMs are neoteric solvents [24] constituted by at least two components, which are able to form a new eutectic phase (which is a liquid below 100°C) by a hydrogen-bond-promoted self-association. Therefore, one of the constituents of these eutectic mixtures must be a hydrogen-bond acceptor (HBA, usually an ionic species such as ChCl or NH_4Cl), whereas the other constituent is

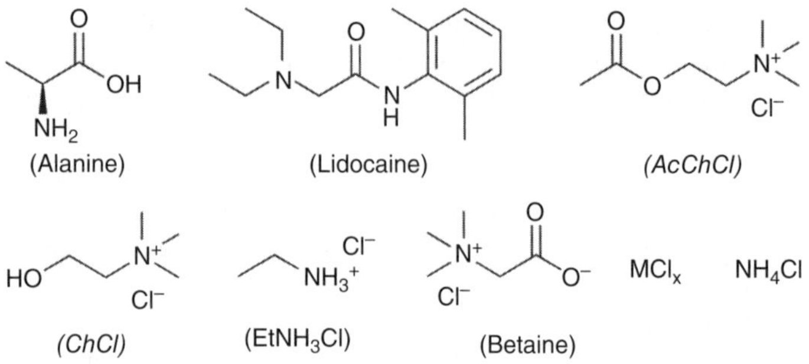

Figure 4.1 Hydrogen-bond acceptors (HBAs) usually employed in the synthesis of DESs and LMMs.

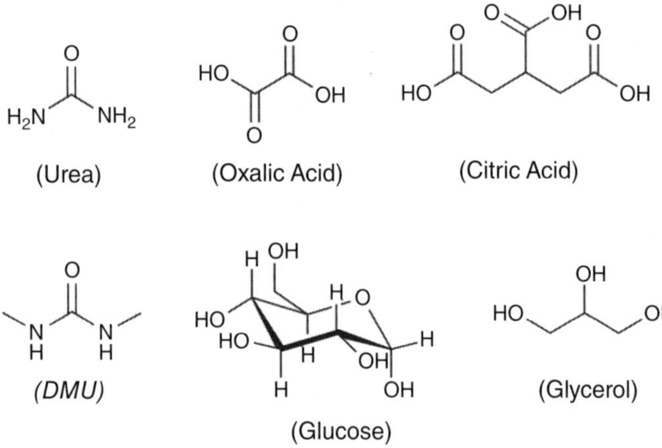

Figure 4.2 Hydrogen-bond donors (HBD) usually employed in the synthesis of DESs and LMMs.

always a hydrogen-bond donor (HBD), like the aforementioned natural polyols (carbohydrates, glycerol), carboxylic acids (lactic acid (Lac), oxalic acid, citric acid) or ureas (Figure 4.2) [20, 21, 25–32].

Bearing in mind the plethora of low-cost, non-toxic, biorenewable and biodegradable organic compounds that are able to generate the aforementioned eutectic mixtures (and compared with the 600 molecular solvents nowadays available), the possibility of creating new green DESs and LMMs for specific tasks is enormous. Moreover, and in contrast to traditional VOC solvents, these eutectic mixtures present a high thermal stability, non-flammability and no vapour pressure, which allows the isolation of the final products through extraction, precipitation or distillation [25–32].

4.3 Deep Eutectic and Low-Melting Mixtures in Metal-Catalysed Organic Reactions

Metal-catalysed organic reactions [33–35] in green solvents (i.e. water [36], ionic liquids [37], glycerol and related biomass-based solvents (see Chapter 1 of this book) [38–40] or sc-CO_2 [41]) have developed into a standard component of the synthetic chemist's toolbox, not only to replace already known organic reaction in VOC solvents, but also to find new ways of perfecting reaction yields, and chemo-, regio- and stereoselectivities [42]. Despite all the recognized advantages associated with the use of non-conventional solvents in organic synthesis, the utilization of green eutectic mixtures (both DESs and LMMs) in metal-catalysed organic reactions is still in its infancy, as the reader will notice in reading this chapter. Thus, in the first section of this chapter we will provide a general overview of metal-catalysed organic reactions in (i) ChCl-based DESs and (ii) LMMs.

4.3.1 Metal-Catalysed Organic Reactions in ChCl-Based Deep Eutectic Solvents

Generally, DESs are composed of a non-toxic quaternary ammonium salt combined with an arbitrary molecule that offers the possibility to function as an HBD. Up to now, the standard choice for a quaternary ammonium salt is the aforementioned choline chloride (ChCl), because it is readily available (as an essential micronutrient [43]), without health-damaging properties [44]. Several ChCl-based eutectic mixtures (containing urea, glycerol or organic acid as HBD) have been used in different chemical applications. For example, the eutectic mixture ChCl/urea (1 : 2) was previously used to: (i) synthesize and crystallize new coordination polymers [45]; (ii) catalyse the fixation of carbon dioxide [46]; and (iii) halogenation reactions [47]. Besides that, ChCl/malic acid can be used for biomass processing [48] and ChCl/glycerol (ChCl/Gly) for sulfur dioxide absorption as well as for the activation of enzymes [49, 50]. However, it was not until the beginning of 2014 that García-Álvarez and co-workers illuminated the way to follow for the employment of ChCl-based eutectic mixtures as green solvents in metal-catalysed organic reactions. The authors studied the catalytic activity of bis(allyl)-ruthenium(IV) derivatives in the redox isomerization of allylic alcohols into saturated carbonyl compounds (Scheme 4.1), including the dinuclear [{Ru(η^3:η^3-$C_{10}H_{16}$)(μ-Cl)Cl}$_2$] (**1**, $C_{10}H_{16}$ = 2,7-dimethylocta-2,6-dien-1,8-diyl) and mononuclear [Ru(η^3:η^3-$C_{10}H_{16}$)Cl$_2$(κ^1-N-benzimidazole)] (**2**) complexes, in different eutectic mixtures [51]. Complex **2** was found to be the most efficient catalyst for the isomerization of a family of allylic alcohols (turnover frequency (TOF) values up to 2970 h^{-1}) in the eutectic mixture ChCl/Gly (1 : 2) under N_2 atmosphere, at 75°C and in the absence of any base as co-catalyst. These optimized reaction conditions allowed the recycling of the catalytic system for

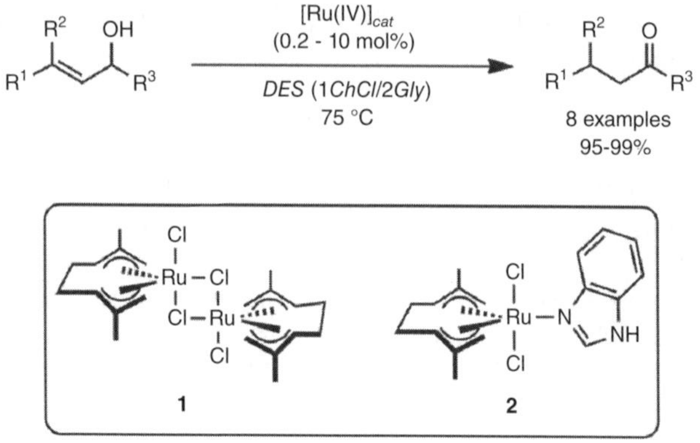

Scheme 4.1 Ru(IV)-catalysed isomerization of allylic alcohols into saturated carbonyl compounds in ChCl-based eutectic mixtures.

four consecutive cycles, with no significant observation of loss of activity during the first two runs. Finally, it should be mentioned that the modification of the ChCl/Gly ratio or the substitution of the glycerol (HBD) by urea or lactic acid in the eutectic mixture considerably decreased the reaction rate.

Later in the same year, García-Álvarez et al. laid their focus on the Au(I)-catalysed cycloisomerizations of unsaturated organic substrates using ChCl-based eutectic mixtures as green solvents. In this sense, the synthesis of heterocycles through an intramolecular addition of a heteroatom-based nucleophile to an alkyne, that is activated by an organometallic π-acid (as defined by Fürstner and Davies [52], i.e. cycloisomerization of γ-alkynoic acids or (Z)-enynols), is particularly interesting. Thus, the metal-catalysed cycloisomerization of γ-alkynoic acids has evolved as a highly attractive methodology for the construction of enol-lactones, densely functionalized heterocycles which are useful synthetic precursors and intermediates. The authors demonstrated that such a process can be selectively performed in the eutectic mixture ChCl/urea (1 : 2) at room temperature and in the presence of air using the iminophosphorane gold(I) compound [AuCl{κ^1-S-(PTA)=NP(=S)(OPh)$_2$}] (**3**, PTA = 1,3,5-triaza-7-phosphaadamantane) as catalyst (Scheme 4.2) [53]. Under these reaction conditions, the catalytic system can be efficiently recycled (up to four consecutive runs).

After this seminal study, the related metal-catalysed cycloisomerization of readily available (Z)-enynols, which represents an attractive and original strategy to produce substituted furans, was successfully performed in different eutectic mixtures with the aid of a new Au(I) complex containing a bis(iminophosphorane) ligand (**4**, Scheme 4.3). The best results in terms of activity and selectivity were obtained with the eutectic mixture ChCl/Gly (1 : 2) at room temperature, in the

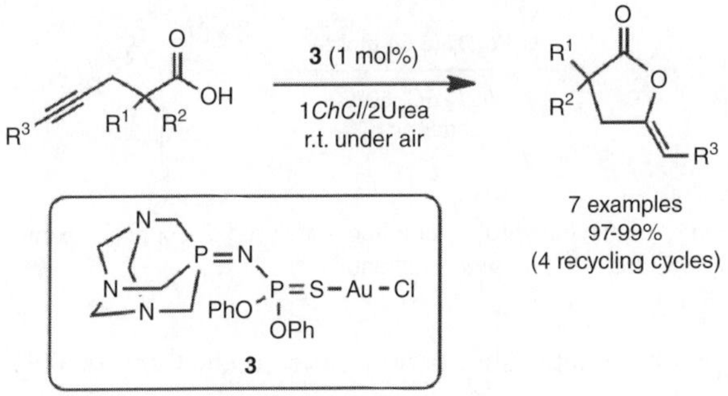

Scheme 4.2 Au(I)-catalysed cycloisomerization of γ-alkynoic acids into saturated enol-lactones in ChCl-based eutectic mixtures.

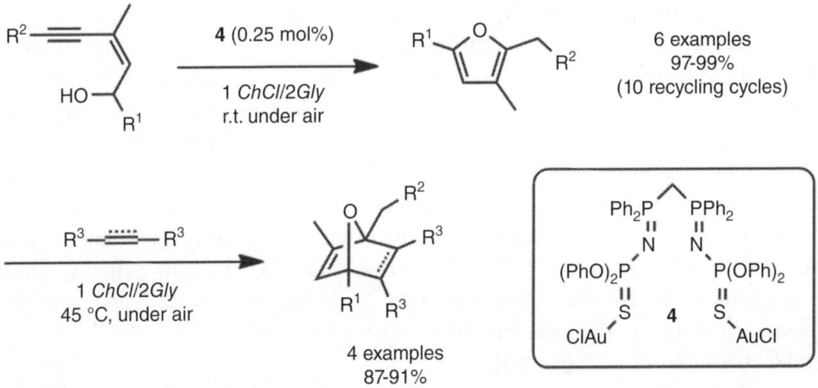

Scheme 4.3 Au(I)-catalysed cycloisomerization of (Z)-enynols and one-pot tandem cycloisomerization/Diels–Alder cycloaddition in ChCl-based eutectic mixtures.

presence of air and in the absence of chloride abstractor [54]. Moreover, the tandem process involving the aforementioned Au(I)-catalysed cycloisomerization of (Z)-enynols and the concomitant Diels–Alder cycloaddition with activated alkynes or alkynes in DESs was employed for the synthesis of 7-oxanorbornadienes and 7-oxanorbornenes, respectively (Scheme 4.3).

Not only transition-metal complexes but also magnetic nanoparticles have been employed as catalyst in organic synthesis in ChCl-based eutectic mixtures. In this regard, Ramón and co-workers [55] have recently reported the synthesis of different tetrahydroisoquinolines using ChCl/ethylene glycol (EG) as green solvent by employing copper(II) oxide impregnated on magnetite as recoverable catalyst (Scheme 4.4). The authors pointed out that the presence of the DES in the reaction medium is crucial, as in its absence the reaction did not take place. For the

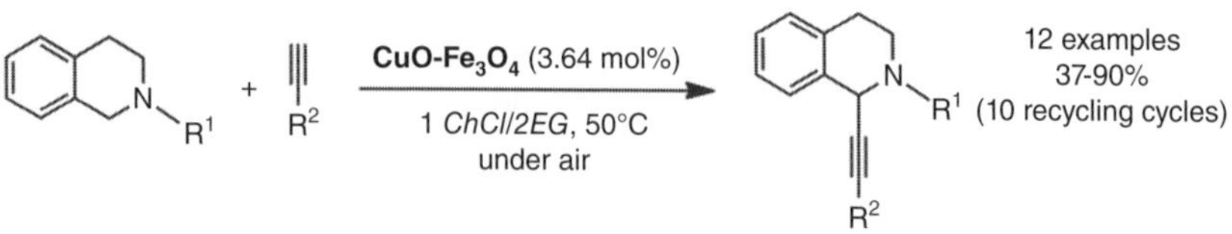

Scheme 4.4 Synthesis of tetrahydroisoquinolines catalysed by copper(II) oxide impregnated on magnetite using ChCl/EG (1 : 2) as green solvent.

first time, a rational relationship between the conductivity of the eutectic mixture and the yield obtained in the reaction was proposed. It is also important to note that the magnetic separation of the catalyst allowed the isolation of the final tetrahydroisoquinolines without using organic solvents, making the process more green and sustainable. In a similar way, Azizi *et al.* have employed closely related magnetically recoverable Fe_3O_4 nanoparticles as catalyst in the eutectic mixture ChCl/urea for: (i) the synthesis of imidazoles [56]; and (ii) the highly efficient cyanosilylation of various aldehydes and epoxides [57]. Finally, titania-supported gold nanoparticles have been successfully applied by Carrier *et al.* as catalyst for the hydrogenation of butadiene in the eutectic mixture ChCl/urea [58].

To end this section, we should point out that several ChCl-based eutectic mixtures containing metallic salts as components (ChCl/MX_n) have been employed as both solvents and catalysts in the following organic procedures: (i) isomerization of aldoximes into primary amides (ChCl/$ZnCl_2$ (1 : 2)) [59]; (ii) palladium-free Sonogashira-type cross-coupling reaction (ChCl/CuCl (1 : 1)) [60]; and (iii) synthesis of indoles, imidazoles and dihydropyrimidin-2-ones (ChCl/$ZnCl_2$ (1 : 2) and ChCl/$SnCl_2$ (1 : 2)) [61–63].

4.3.2   Metal-Catalysed Organic Reactions in Low-Melting Mixtures

Low melting eutectic mixtures of sugar, urea and salt (also known in the literature as 'sugar melts' or 'sweet melts') were introduced by König and co-workers in 2005 as environmentally benign solvents, since they are safe, non-toxic, biodegradable and easily available from renewable feedstocks [23]. In this pioneering work, these stable eutectic mixtures (with melting points above 60°C) were successfully applied as solvents for Diels–Alder cycloadditions. One of the most important advantages of this procedure is related with the work-up of the LMMs, which can be conducted by: (i) simple addition of water (leading to phase separation); or (ii) distillation of low-boiling-point organic products from the reaction mixture.

After this pivotal study, a variety of metal-catalysed organic reactions have been developed by König *et al.* in different sugar-based LMMs. Firstly, the authors

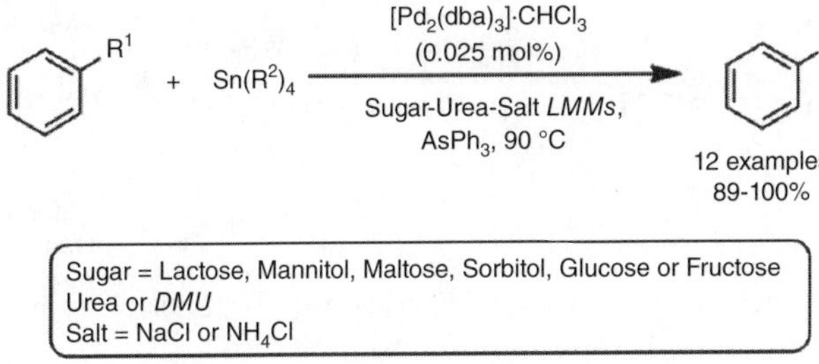

$$[Pd_2(dba)_3] \cdot CHCl_3$$
$$(0.025 \text{ mol\%})$$

Sugar-Urea-Salt *LMMs*,
$AsPh_3$, 90 °C

12 examples
89–100%

Sugar = Lactose, Mannitol, Maltose, Sorbitol, Glucose or Fructose
Urea or *DMU*
Salt = NaCl or NH_4Cl

Scheme 4.5 Pd(0)-catalysed Stille C–C coupling reaction in sugar-based LMMs.

reported the fast and efficient Pd-catalysed allyl transfer of tetralkyltin reagents (Stille reaction, Scheme 4.5) [64]. The smooth formation of the desired aromatic products was attributed to the high polarity and nucleophilic character of the sugar-based LMMs. Under these reaction conditions, not only Stille alkylations but also biaryl synthesis can be efficiently performed, as the catalyst loading can be reduced to 0.001 mol.% and the catalytic system can be recycled up to three consecutive runs. Finally, it is important to note that König and co-workers have almost dominated the field of Pd-catalysed cross-coupling reactions in sugar-based LMMs by applying this new green procedure to Suzuki [65], Heck [66] and Sonogashira [66] reactions.

Apart from sugars in the 'sweet' LMMs, other renewable materials are suitable to form LMMs in combination with different ureas. In this sense and for comparison, König's group also studied several metal-catalysed organic reactions in the acid-based LMM citric acid/dimethylurea (DMU), such as: (i) the aforementioned Pd-catalysed Stille coupling [64]; and (ii) the catalytic hydrogenation of methyl α-cinnamate using Wilkinson's complex [RhCl(PPh$_3$)] as catalyst. In this Rh-catalysed hydrogenation procedure, the reaction progresses rapidly, cleanly and quantitatively in the citric acid/DMU melt and less efficiently (50% conversion) in other sugar-based LMMs (containing mannitol and sorbitol, respectively) [65]. More recently, this acid-containing eutectic mixture (citric acid/DMU) was utilized by Zhang and co-workers in their investigations of the catalytic activity of $CuFeO_2$ nanoparticles in the one-pot three-component reaction of 2-aminopyridines, aldehydes and terminal alkynes for the synthesis of an array of imidazole[1,2-*a*]pyridines (Scheme 4.6) [67]. The authors examined this reaction in a variety of both organic solvents (such as methanol, ethanol, toluene, acetonitrile, *N,N*-dimethylformamide) or eutectic mixtures (fructose/DMU, mannose/DMU/NH$_4$Cl and lactose/DMU/NH$_4$Cl), and it was found that the reaction in citric acid/DMU gave the best result. After the reaction was accomplished, the citric acid/DMU eutectic mixture as well as the magnetic

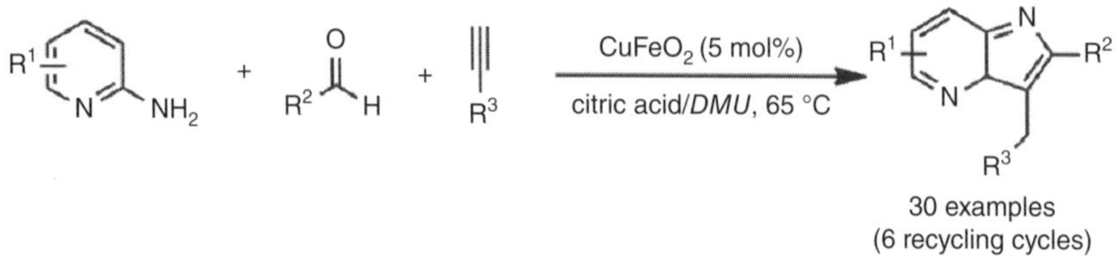

Scheme 4.6   CuFeO$_2$-catalysed synthesis of imidazole[1,2-a]pyridines in the LMM citric acid/DMU.

nanoparticles could be recovered and reused in six consecutive runs, thus making this protocol economically and potentially viable.

L-Carnitine (a natural betaine) has also been employed as a biorenewable, non-toxic and stable component for the synthesis of the corresponding LMM with urea (L-carnitine/urea). Like the previously reported sugar-based LMMs, this betaine-containing eutectic mixture showed good solvent properties in the abovementioned Pd-catalysed Heck coupling [66]. Remarkably, the CuI-catalysed Huisgen cycloaddition of benzyl azide and phenylacetylene (CuAAC) to generate the corresponding triazole was improved in the L-carnitine-based LMM, when compared with sugar-based eutectic mixtures [66]. The authors proposed the *in situ* formation of the complex [CuI·(L-carnitine)], resulting in a stabilizing effect during the reaction.

Finally, β-cyclodextrins (β-CD, a cyclic carbohydrate constituted of seven D-glucopyranose units) were proposed by Jérôme, Tilloy and co-workers as new natural components for the synthesis of a family of LMMs (with melting points around 90°C) in combination with DMU. The authors assessed the applicability of these eutectic mixtures in metal-catalysed organic reactions by studying the hydroformylation of alkenes and the Tsuji–Trost reaction (Scheme 4.7). In both cases, (i) the presence of a water-soluble phosphine (TPPTS = tris(3-sulfophenyl)phosphine trisodium salt) in the catalytic mixture was mandatory, and (ii) best results were obtained when randomly methylated β-cyclodextrins (RAME-β-CD) were used in the synthesis of the eutectic mixture [68, 69].

4.4   Conversion of Carbohydrates

Carbohydrates derived from biomass are one of the most abundant renewable resources and possess a great potential to become raw materials for the production of many high-value products. Hence a plethora of research has been focused on the conversion of carbohydrates to key platform molecules. Among the reactions studied, the synthesis of 5-hydroxymethylfurfural (HMF) is of prime interest, HMF being a key platform chemical for the production of a variety

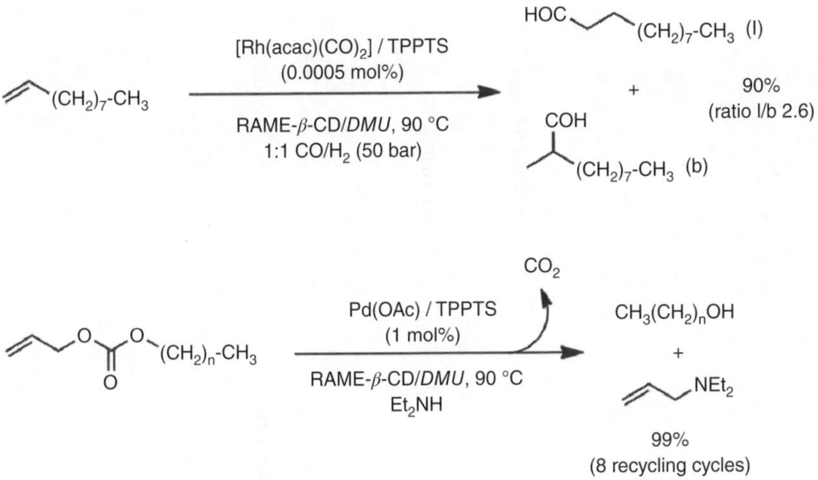

Scheme 4.7 Rh-catalysed hydroformylation of alkenes and Pd-catalysed Tusji–Trost reaction in the eutectic mixture RAME-β-CD/DMU.

of downstream chemicals and biofuels [70–74]. HMF can be obtained from cellulose by successive depolymerization/isomerization/dehydration reactions (Scheme 4.8).

The conversion of carbohydrates (cellulose, glucose, fructose) to HMF is catalysed by an acid catalyst. In this reaction, the nature of the solvent plays a key role since it has to be inert to HMF, to dissolve a large amount of carbohydrates and to avoid secondary reaction of HMF rehydration to formic and levulinic acids. In the literature, various solvents have been used. Among the investigated solvents, we can cite dimethylsulfoxide (DMSO), which leads to an HMF yield greater than 85% [75–78]. However, the extraction of HMF from DMSO is cost-efficient and DMSO can be decomposed into toxic undesired products under acid conditions. Water was also widely used in this reaction, leading to low yield of HMF due to the rehydration of HMF to levulinic acid and formic acid [79–83]. That is the reason why several biphasic media were investigated where water was associated with an organic solvent allowing the dissolution of HMF. Another class of solvents studied in the synthesis of HMF is ionic liquids (ILs) [84–90]. They allow the formation of a high yield of HMF and HMF can be recovered by liquid–liquid extraction. However, the price and the toxicity of ILs have hampered their used at an industrial scale. In this context, LMMs and DESs based mostly on ChCl were investigated in the synthesis of HMF from carbohydrates. The synthesis of furfural from the hemicellulosic part of lignocelluloses was also investigated in the presence of eutectic mixture and the results obtained will be discussed below.

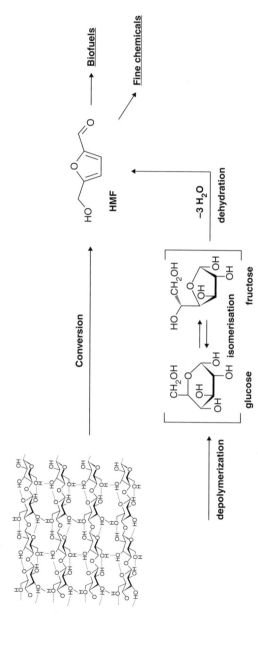

Scheme 4.8 Synthesis of HMF from cellulose.

4.4.1   Synthesis of 5-Hydroxymethylfurfural from Carbohydrates in Low-Melting Mixture

The groups of Han and König were among the first to study the catalytic conversion of fructose to HMF in various LMMs [91, 92]. In a first step of experiments, the authors have studied the formation of HMF from D-fructose in LMM with urea, N,N'-dimethylurea (DMU) or N,N'-tetramethylurea (TMU) in the presence of several catalysts ($ZnCl_2$ and $CrCl_3$). In urea and DMU mixtures, low yields of HMF were observed (below 8%), due to the formation of urea–fructose condensation products. In contrast, in TMU mixture, in which a condensation between TMU and fructose cannot be observed, HMF was obtained in 89% yield at 100°C after 1 h of reaction [92] (Scheme 4.9).

Zhang *et al.* [93] have studied the tetraethylammonium chloride (TEAC)/ fructose LMMs to efficiently catalyse the fructose dehydration solely or by using $NaHSO_4 \cdot H_2O$ as a catalyst. In the absence of $NaHSO_4 \cdot H_2O$ and a fructose to TEAC weight ratio lower than 0.52, a yield of HMF above 80% was observed with a fructose content of 33.3 wt.%. As currently observed, if the fructose concentration was further increased, a gradual decrease in the HMF yield occurred due to the formation of by-products such as insoluble humins or soluble polymers. The authors have explained the catalytic effect of TEAC by an interaction of the $N^+(CH_2CH_3)_4$ cation with fructose leading to its dehydration. However, no proof of this mechanism was reported in this work. The authors have investigated the comparison of the fructose reactivity in different ammonium salts: (i) quaternary ammonium chlorides, i.e. trimethylammonium chloride (TMAC), TEAC, tributylammonium chloride (TBAC), benzyltrimethylammonium chloride (BeTMAC), trimethylphenylammonium chloride (PhTMAC), benzyltriethylammonium chloride (BeTEAC) and ChCl), and (ii) alkylamine hydrochlorides, i.e. trimethylamine hydrochloride (TMHC) and dimethylamine hydrochloride (DMHC). All of these LMMs exhibited a yield of HMF above 65% at 120°C after 70 min of

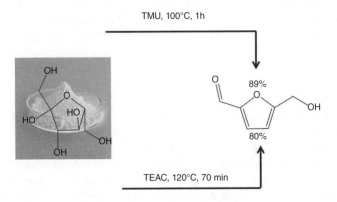

Scheme 4.9 Synthesis of HMF in LMM.

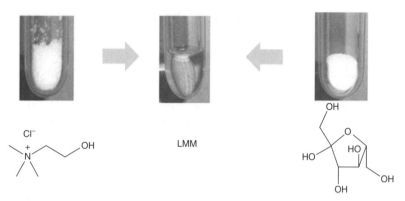

Scheme 4.10 Formation of an LMM between ChCl and fructose. (A colour version of this scheme appears in the plate section.)

reaction. However, only TEAC presented a yield higher than 80% (Scheme 4.9). If NaHSO$_4$·H$_2$O was added as a catalyst to the fructose/TEAC system, a yield of 73% of HMF from 60 wt.% of fructose was obtained. This result shows the beneficial effect of LMM solvents. Hence, a high substrate concentration is usually preferred to improve the efficacy of the process. However, most of the systems described in the literature in water, polar solvent or ILs are only tolerant towards low to moderate fructose concentrations (no more than 30 wt.%). The use of LMM allows the utilization of high concentration of the reactant. Another study leading to the same conclusion will be reported later on. However, using these solvents, the conversion of glucose was unsuccessful with or without the addition of NaHSO$_4$·H$_2$O to the fructose/TEAC system. The extraction of HMF from the reaction medium was conducted by adding THF in the TEAC/fructose/NaHSO$_4$·H$_2$O system.

König *et al.* have also carried out the synthesis of HMF from mixtures of ChCl/fructose (Scheme 4.10), ChCl/glucose, ChCl/sucrose and ChCl/inulin [92]. Various acid catalysts (Amberlyst 15, FeCl$_3$, ZnCl$_2$, CrCl$_2$, CrCl$_3$, *p*-toluenesulfonic acid (pTSOH), scandium triflate (Sc(OTf)$_3$) and montmorillonite) were investigated, and it was shown that, in the presence of a mixture of fructose and ChCl (4 : 6), 59% of HMF was obtained at 100°C after 0.5 h in the presence of pTSOH.

As currently described, the authors have observed significant amounts of black humins upon prolonged reaction time. In the conversion of glucose to HMF, CrCl$_2$ was the most active and selective catalyst, leading to 45% of HMF in a mixture of glucose and ChCl (4 : 6) at 100°C. These LMMs can be efficient for the conversion of more complex carbohydrates. Hence, the direct conversion of carbohydrates such as sucrose and inulin to HMF going through a tandem reaction of hydrolysis/dehydration was investigated. The authors have shown that 62% of HMF was obtained from sucrose after 1 h of reaction at 100°C and 57% of HMF was obtained from inulin in the presence of pTSOH at 90°C for 1 h. Prolonged reaction time or temperature higher than 100°C led to the formation of insoluble humins.

Liu *et al.* have studied the conversion of glucose to HMF in the presence of various Lewis acid catalysts using ChCl [94]. The best HMF yield (70%) was achieved in a mixture of $ChCl/H_2O$ (1 : 1) at 90% of glucose conversion at 165°C in the presence of 3 mol.% of $AlCl_3 \cdot 6H_2O$ after 3 h of reaction, using methyl isobutyl ketone (MIBK) for the extraction of HMF. The role of water was also studied and they have shown that, when the $ChCl/H_2O$ ratio was increased from 0 to 1, a gradual increase of the HMF yield was observed up to 70%, attributed to the beneficial effect of ChCl on the reaction selectivity (Scheme 4.11). One should mention that, in this concentration range, the partition coefficient of HMF between the aqueous and MIBK phases did not vary (1.2). When the $ChCl/H_2O$ ratio was higher than 1, the HMF yield decreased down to 50% due to the side rehydration reaction, and the partition coefficient of HMF dropped, highlighting the strong interaction of ChCl with the produced HMF. The conversion of cellulose was also investigated in a $ChCl/H_2O$ ratio of 1 using MIBK as extracted solvent; 27% of HMF was produced. When cellulose was ball-milled, an HMF yield of 46% was achieved in the presence of $FeCl_3$ and $AlCl_3$ as catalysts. In the conversion of cellulose to HMF, three elementary steps are required to convert cellulose to HMF. Thus, the average yield for each step is approximately 80%, showing the good selectivity of this process.

Based on this study, the same authors have investigated the heterogeneously catalysed conversion of hexoses to HMF in low-boiling-point organic solvents,

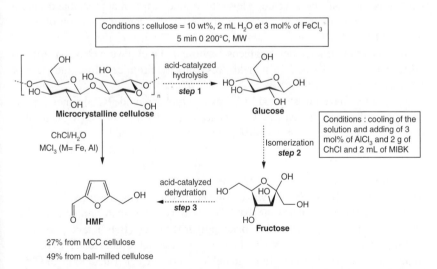

Scheme 4.11   Conversion of microcrystalline cellulose to HMF in the presence of $ChCl/H_2O/$ MIBK solvent.

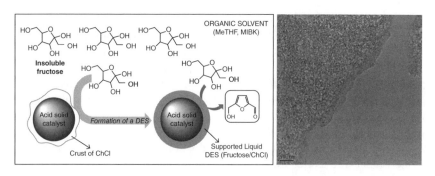

Scheme 4.12   Conversion of fructose to HMF in organic solvents using a supported LMM.

such as methyltetrahydrofuran (MeTHF) and MIBK [95] (Scheme 4.12). Hence, the use of these organic solvents helps to prevent the side rehydration of HMF in water and the recovery of the solvent and the catalyst is favoured. However, carbohydrates are not soluble in these organic solvents. To overcome this problem, the coating of a solid acid catalyst ($AlCl_3@SiO_2$) with ChCl was investigated. This supported LMM allowed the *in situ* formation of an LMM with hexoses on the catalyst surface, thus facilitating the contact between the two suspensions of silica and carbohydrates and resulting in the formation of 51% of HMF in MeTHF (in the presence of 39 wt.% of ChCl on $AlCl_3@SiO_2$) and 65% of HMF in MIBK (in the presence of 17 wt.% of ChCl on $AlCl_3@SiO_2$). They have shown that, owing to the insolubility of ChCl in MeTHF or MIBK, the $AlCl_3@SiO_2$ coated with ChCl was successfully recycled at least three times without appreciable decrease of activity. More importantly, because of the use of a catalytic amount of ChCl, the MeTHF or MIBK phase contains more than 95 wt.% of the produced HMF. Different carbohydrates such as glucose, xylose and inulin were also converted to HMF with 38%, 43% and 45% yields, respectively. From cellulose, the yield of HMF remained lower than 10%.

Prasad *et al.* have investigated LMM using graphene oxide (GO) nanosheets as an acid catalyst and the reaction was carried out under microwave (MW) irradiation [96]. The extraction of the HMF produced was performed using ethyl acetate as an organic solvent. If ethyl acetate was used alone, no HMF was observed. However, 76% of HMF was produced from the dehydration of 2.5% w/v fructose in the presence of 0.1% w/v of GO, 1% w/v of ChCl at 100°C for 30 min under MW irradiation. If the reaction was carried out in ChCl in the absence of GO, 22% of HMF was obtained, indicating that GO acts as a catalyst in this process. If ChCl was replaced by another additive, betaine hydrochloride (BHC), a co-product of the sugar beet industry, an HMF yield up to 70% was achieved. Hence, BHC is an ionic, safe, cheap (€3/kg) and biodegradable carboxylic acid (metabolite of ChCl) with a pK_a of 1.9. The authors have shown that, in the presence of GO additives (BHC or ChCl), the HMF yield was highest when fructose was used as substrate, followed by glucose, sucrose, mannose and galactose. Moreover, the solvent was recycled and successfully reused. During the course of reaction, GO was reduced to produce six-layered graphene nanosheets (96% recovery).

In 2012, de Oliveira Vigier *et al.* [97] reported that BHC in combination with ChCl and water can act as an acid catalyst for the production of HMF from fructose and inulin (Scheme 4.13). In a ChCl/BHC/water (10 : 0.5 : 2) mixture, HMF was produced with 63% yield (at 130°C from 40 wt.% of fructose). If the reaction was performed in a biphasic system using MIBK as an extraction solvent, HMF was recovered with a purity higher than 95% (isolated yield of 84% from 10 wt.% of fructose). Moreover, the ChCl/BHC/water system was successfully recycled up to seven times.

Later, Liu *et al.* have shown that, under a pressure of CO_2, fructose and inulin can be converted to HMF in a ChCl/fructose LMM [98]. As previously shown by Han (Scheme 4.13) [99], with water, CO_2 can form carbonic acid that has a pK_a low enough to catalyse the dehydration of fructose to HMF. Hence, CO_2 initially reacted with water contained in fructose and in the DES, resulting in the formation of carbonic acid in a sufficient amount to initiate the dehydration of fructose to HMF. After 90 min of reaction at 120°C under 4 MPa of CO_2, a yield of 74% of HMF was obtained from 20 wt.% of fructose. HMF was extracted using MIBK with a purity of 98%. Remarkably, the present system was tolerant to very high loading of fructose, whereas most of the reported solvents suffer from a low selectivity to HMF when the fructose loading was higher than 30 wt.%. For instance, fructose with a loading of 100 wt.% was successfully dehydrated to HMF in a ChCl/CO_2 system without appreciable decrease of the HMF yield (66%). The authors have ascribed the tolerance of this system to a high loading of fructose to the strong interaction between produced HMF and ChCl resulting in the stabilization of HMF in the reaction medium. As shown in Scheme 4.13, when neat HMF and ChCl were mixed together, a melt was readily obtained at a fructose content higher than 60 wt.%. Such a melt might be responsible for the surprising stability of HMF in this system when using a high loading of fructose. It is indeed known that when a chemical is engaged in the formation of a eutectic mixture its reactivity is drastically reduced.

Another study was performed on the use of a series of acid–base heteropolyacid (HPA) catalysts with the formula $(C_6H_{15}O_2N_2)_{3x}H_xPW_{12}O_{40}$ (abbreviated as $Ly_{3x}H_xPW$) – containing double NH_2 groups from amino acids as base functional groups and an $H_3PW_{12}O_{40}$ fragment as the acid one – in the production of HMF from fructose in the ChCl solvent [100]. Up to 92% yield of HMF was obtained in the ChCl–Ly_2HPW system within 1 min. It is reported that the rate-determining step of fructose dehydration is the enolization of fructose. Ly_2HPW could provide the concerted activation of fructose at the C–OH bond of the anomeric cation by the proton (electrophile) and at the C–H of C1 of the fructose via hydrogen bonding by N (nucleophile) (Scheme 4.14) as well. This demonstrated that the coexistence of hydrogen-bond donors (–NH_2) and protons in Ly_2HPW gave

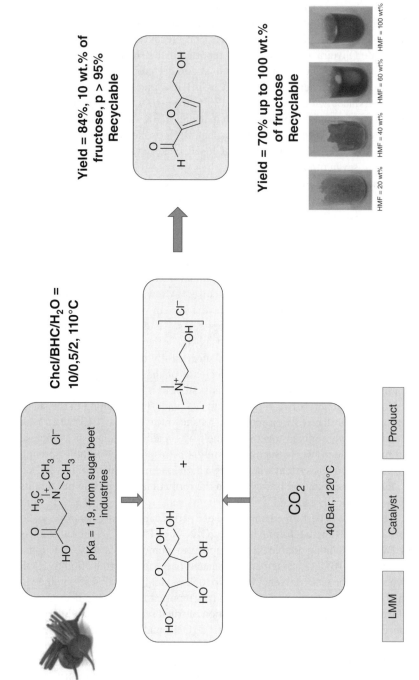

Scheme 4.13 Acid-catalysed dehydration of fructose in ChCl-derived LMM. (A colour version of this scheme appears in the plate section.)

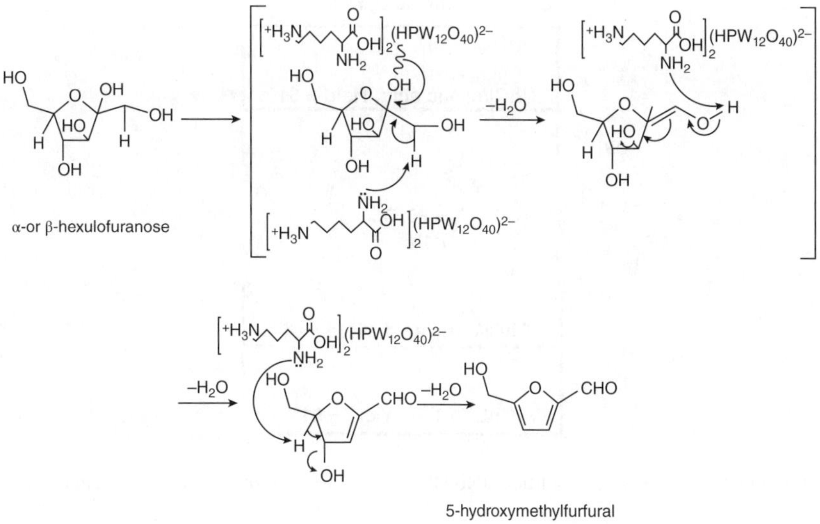

5-hydroxymethylfurfural

Scheme 4.14 Mechanism for the dehydration of fructose to HMF in ChCl catalysed by acid–base HPA catalyst [100]. From Zhao *et al.* (2014) *RSC Adv.*, **4**, 63055–63061. Reproduced by permission of RSC.

synergistic potentials for the selective and efficient conversion of fructose to HMF. Therefore, high efficiency (low temperature and short time) and high yield of HMF were obtained catalysed by double acid–base HPA Ly_2HPW in ChCl solvent.

4.4.2 Synthesis of Furanic Compounds (Furfural and 5-Hydroxymethylfurfural) in ChCl-Based Deep Eutectic Solvents

Han and co-workers have studied the synthesis of HMF in the presence of a ChCl-based mixture containing carboxylic acids [91]. At 80°C for 1 h, above 90% of fructose was converted with 45%, 61% and 84% of selectivity to HMF in ChCl/malonic acid, ChCl/oxalic acid and ChCl/citric acid monohydrate, respectively (Scheme 4.15). HMF was found to be relatively stable at this temperature in a mixture of ChCl/citric acid monohydrate. These results show that the synthesis of HMF from fructose can also be carried out in DESs. If the reaction was performed in a biphasic medium using ethyl acetate to extract HMF, 87% HMF yield was observed in the presence of ChCl/citric acid monohydrate phase that was easily recyclable. They have also studied the one-pot conversion of inulin and 56% and 51% HMF yields respectively were obtained in ChCl/oxalic acid or ChCl/citric acid [101]. The yield of HMF in the ethyl acetate/ChCl/oxalic acid biphasic system was higher (64% vs. 56%) than in the absence of ethyl acetate. The DES phase was recycled directly after removing the ethyl acetate

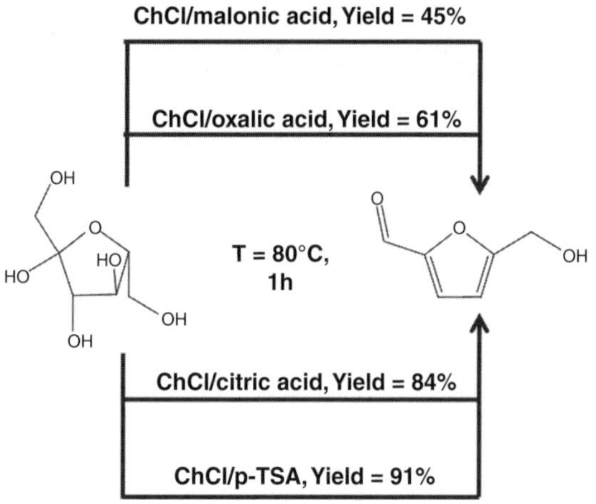

ChCl/malonic acid, Yield = 45%

ChCl/oxalic acid, Yield = 61%

T = 80°C,
1h

ChCl/citric acid, Yield = 84%

ChCl/p-TSA, Yield = 91%

Scheme 4.15   Conversion of fructose to HMF in the presence of ChCl/carboxylic acid DES.

phase containing HMF. Up to six cycles could be performed without a significant loss in the HMF yield.

Recently Al-Duri and co-workers [102] have studied the dehydration of fructose to HMF in a DES composed of ChCl and *p*-toluenesulfonic acid monohydrate (*p*-TSA), which obviates the addition of an acid catalyst. The selected DES was found to efficiently convert fructose to HMF, a 91% yield of HMF being obtained at a feed ratio of 2.5 wt.%, DES mixing molar ratio of 1 : 1, 80°C temperature and 1 h reaction time (Scheme 4.15). The use of *p*-TSA and the use of carboxylic acids as previously mentioned as hydrogen-bond donor and catalyst reduce the processing cost. However, the separation of HMF from the ChCl/*p*-TSA was not evaluated in this study.

Matsumiya and Hara have conducted the conversion of glucose to HMF in the presence of DES made of choline salts and carboxylic acids (succinic acid, malic acid, glycolic acid, mandelic acid, salicylic acid, oxalic acid, benzoic acid) or urea [103]. In this study, boric acid $B(OH)_3$ was added as a promoter to isomerize glucose to fructose before the dehydration of fructose to HMF. Unfortunately the HMF yields remained very low (below 5%) due to the formation of HCl in the reaction media leading to condensation products such as humins when the reaction was carried out at 140°C for 2 h in the presence of boric acid. To overcome this problem, ChCl was replaced by choline dihydrogen citrate, another inexpensive and commercially available choline salt. In this case, the mixtures were melted even at room temperature, except for benzoic acid. The highest yield to HMF (42%) was obtained using glycolic acid/ChCl in the presence of $B(OH)_3$. The role of the boric acid was investigated. If boric acid was replaced by $CrCl_3$ or $CrCl_2$, no HMF was produced in similar conditions. The authors explained that the isomerization pathway from glucose to fructose goes through the complexation with $B(OH)_3$, the role of $B(OH)_3$ being dual: (i) to decrease the energy barrier for the formation of the intermediate species; and (ii) to increase the exothermicity for the overall isomerization. One can note that the decrease in the HMF yield by large amounts of $B(OH)_3$ suggests that the excessively strong complexation might inhibit the further reactions. If solely neat choline dihydrogen citrate was used in the presence of $B(OH)_3$, the HMF yield was very low (7%), showing that carboxylic acids are required for the production of HMF since they are involved in the dehydration of fructose to HMF. The authors have studied the effect of the strength of the acid by replacing carboxylic acids by the strong acid, methanesulfonic acid. They have demonstrated that the strength of the acid should be low to produce HMF in order to avoid secondary reactions. The effect of a co-solvent to reduce the viscosity of the molten mixture was studied, and the authors have shown that the addition of water was beneficial for the formation of HMF, a yield of 60% being obtained at 140°C for 4h by avoiding the formation of humins. However, no explanation was given in this study.

Another study aims at optimizing the conversion of native cellulose and ligno-cellulosic residues into HMF and furfural (coming from the hemicellulose part of lignocellulosic biomass) in the presence of several eutectic mixtures, namely ChCl/urea, ChCl/oxalic acid, ChCl/betaine and TEAC at 170°C for 2 min [104]. In this study the effect of the catalyst, the nature of the eutectic mixture and the water content were investigated. The authors have shown that, depending on the eutectic mixtures, the catalyst and the diluent are different. Hence, the ChCl/betaine eutectic mixture afforded HMF and furfural yields of 23% and 4%, respectively, in the presence of KOH, without organic diluents. The ChCl/urea mixture led to very low yields in the conversion of native cellulose (below 2%), and the same happened to TEAC when it operated with basic oxides. In contrast, the use of a tungstic acid in the presence of TEAC improved the yield (14%). A similar trend was observed for the ChCl/betaine mixture, when it was used in combination with sodium molybdate salt, leading to 16% of HMF and 8% of furfural. HMF and furfural yields of 22% and 10%, respectively, were obtained in ChCl/oxalic acid mixture in the presence of TiO_2 (which is an amphoteric oxide) catalyst and tri-ethylene glycol as diluent (Scheme 4.16). The highest HMF plus furfural yield was observed for this combination, with a value of 32%, which indicates that the density of acidic and basic sites is important for the breakdown of cellulose chains and for the dehydration of glucose into HMF. The kinetic studies conducted for the separate production of HMF and furfural for the different substrates under study have demonstrated that the highest HMF production yields are obtained at 200°C, whereas the highest furfural productions occur at 140°C. Finally, when stirring is combined with microwave treatment, and when an additional 15 min ultrasonic pre-treatment is also conducted, the HMF plus furfural production yields are dramatically increased (vs. regular microwave treatment without stirring).

ChCl/citric acid DES containing $AlCl_3·6H_2O$ was used to produce furfural from xylan or xylose by Zhang and Yu [105]. The authors have shown that a synergistic

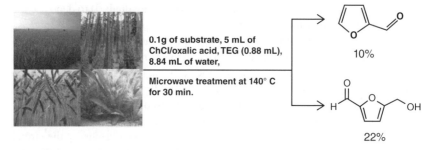

0.1g of substrate, 5 mL of
ChCl/oxalic acid, TEG (0.88 mL),
8.84 mL of water,

Microwave treatment at 140° C
for 30 min.

10%

22%

Scheme 4.16   Conversion of lignocellulosic biomass to HMF and furfural in DES. (A colour version of this scheme appears in the plate section.)

effect between the Brønsted and Lewis acids led to the conversion of xylan and xylose to 59% and 54%, respectively, of furfural at 140°C after 15 and 25 min of reaction. $AlCl_3 \cdot 6H_2O$ is supposed to catalyse the isomerization of xylose to xylulose, while citric acid promoted the dehydration of xylulose to furfural. As previously shown in various studies, if the reaction is carried out in a biphasic system (DES/MIBK), the furfural yield was increased to 73% and the DES phase containing $AlCl_3 \cdot 6H_2O$ was recycled at least five times. At temperatures higher than 140°C, the recycling was hampered by the degradation of the DES phase. In a similar study, Liu *et al.* [106] have shown that at 150°C up to 70% of furfural can be obtained from xylan and wheat straw in the presence of BHC/water mixture. In this study the proof of the formation of a eutectic mixture between BHC and furfural was not evidenced. However, it was found that BHC was also capable of strongly interacting with furfural, resulting in high yields of furfural.

4.5   Extraction with or from Deep Eutectic Solvents

As described previously, HMF can be produced in LMM or DES due to its high solubility in these media. Based on these results, Kobayashi *et al.* have studied the extractive recovery of HMF from such a reaction medium [107]. The reaction was carried out in the presence of a DES composed of ChCl and citric acid. The partition coefficient of HMF in a biphasic system was investigated. Several organic solvents such as THF, ethyl acetate, 2-butanol, MIBK and toluene were studied. When THF was used as the organic phase, the highest partition coefficient of 0.93 was obtained. In addition, when 2-butanol, ethyl acetate, MIBK and toluene were used as the organic solvent with ChCl/citric acid, partition coefficients of 0.91, 0.46, 0.45 and 0.05, respectively, were obtained. Despite the high partition coefficient of THF and 2-butanol, these solvents exhibit some drawbacks. Hence, THF often forms peroxides that would cause explosion during the concentration and reuse of the solvent. In the case of 2-butanol, it easily mixed with ChCl/citric acid, which would cause a loss of this DES and would complicate operations during the

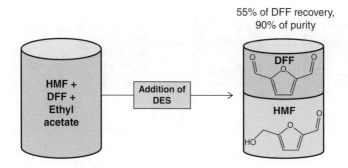

Scheme 4.17 Extraction of DFF using DES.

purification of HMF. The authors did not observe these issues when ethyl acetate or MIBK were used. In this study, it was shown that HMF can be extracted from a eutectic mixture with an appropriate organic solvent. This is the key point in the utilization of DES or LMM in reactions.

The HMF can be converted into different products such as esters. Hence, Dominguez de Maria and co-authors investigated the biocatalytic (trans)esterification of HMF with different acyl donors (ethyl acetate, ethyl hexanoate, dimethyl carbonate, soybean oil, propionic acid, hexanoic acid and lauric acid) [108]. Under solvent-free conditions, this reaction affords HMF esters with 80% yield after 24 h of reaction at 408°C. However, despite the high selectivity of the reaction, the separation of unreacted HMF from the HMF esters should be performed. To this aim, ChCl-based DESs (composed of ChCl and either glycerol or xylitol or urea) were used for the selective extraction of HMF from HMF esters. More than 90% of HMF esters, with purity higher than 99%, was isolated after selective extraction of residual HMF by the ChCl-based DES. According to the nature of the DES, the separation was always very selective and the optimal DES/reaction mixture volume ratio was found to be 1 : 1.

Another product that can be obtained from HMF is diformylfuran (DFF). The conversion of HMF to DFF is carried out in the presence of oxidative catalyst. Li *et al.* have used an enzyme (alcohol oxidases or galactose oxidase) to carry out this reaction [109]. They have shown that a quantitative yield of DFF (92%) was obtained with 8 U galactose oxidase in the presence of 1.1 mg of catalase and 1.3 mg of horseradish peroxidase after reaction for 96 h in 2 ml deionized water at 25°C. The separation of the residual HMF and the formed DFF was investigated using DES including ChCl/glycerol (1 : 2), ChCl/urea (1 : 2) and ChCl/xylitol (1 : 1) to extract HMF from ethyl acetate containing DFF and HMF. The authors have shown that for a ChCl/glycerol DES/ethyl acetate mixture of 5 : 1 (v/v) ratio, DFF can be extracted with 55% of recovery and a purity higher than 90% (Scheme 4.17). This study is a proof of concept that DES can be used to extract DFF from a mixture of HMF and DFF, showing the high interest of using DES.

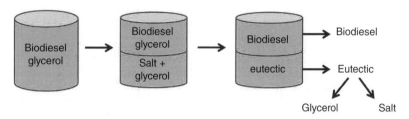

Scheme 4.18   Removal of excess glycerol in biodiesel production using a eutectic salt/glycerol mixture.

DES was also extensively studied for the extraction of glycerol from biodiesel (obtained through methanolysis of triglycerides) (Scheme 4.18). Abbott *et al.* have conducted this extraction by adding a basic quaternary ammonium salt capable of producing *in situ* a DES with residual glycerol [110]. Glycerol was efficiently extracted using a mixture of a glycerol/salt molar ratio of 1 : 1 by using ethylammonium chloride and 1-chloro-*N,N,N*-trimethylmethanaminium chloride as salts. If acetylcholine chloride, choline chloride and tetrapropylammonium bromide were used, the extraction of glycerol was less effective. Shahbaz *et al.* have shown that glycerol can be extracted from palm-oil-based biodiesel using various DESs based on the methyltriphenylphosphonium bromide [111]. The authors have explained that the improvement in the glycerol extraction was due to the high affinity of DESs with glycerol based on hydrogen bonding. DESs based on methyltriphenylphosphonium bromide as salts and glycerine, ethylene glycol and triethylene glycol as hydrogen-bond donor were not highly efficient to remove residual glycerine contained in biodiesel. Only DESs composed of ChCl/glycerol (1 : 2 molar ratio), while respecting the DES/biodiesel molar ratio of 2 : 1, 2.5 : 1 and 3 : 1, were found to be efficient. However, DESs made of ethylene glycol or triethylene glycol were found to be more efficient in removing residual glycerol from biodiesel. The optimum DES/biodiesel molar ratio using ethylene glycol or triethylene glycol was 0.75 : 1. However, no investigation on the recovery and the recyclability of the DES was carried out. The authors have also demonstrated that the residual catalyst KOH used in the transesterification of oils can be removed from the reaction media using DES based on choline chloride or methyltriphenylphosphonium bromide (MTPB) salts [112]. In such a case, glycerol, ethylene glycol 2,2,2-trifluoroacetamide and triethylene glycol were used as hydrogen-bond donors. An increase of the DES/biodiesel and ChCl/HBD led to a higher KOH extraction efficiency. For instance, the ChCl/glycerol and MTPB/glycerol DESs allowed removal of 98.5% and 94.6%, respectively, of KOH from palm-oil-based biodiesel.

The high viscosity and solid state of most eutectic mixtures at room temperature restrict their application as extraction solvents. Hence, Choi and co-workers proposed to decrease the viscosity of some LMMs such as sucrose/ChCl, fructose/glucose/sucrose, proline/maleic acid and glucose/ChCl by adding water [113]. The optimal water content was investigated for each LMM used.

For instance, the extraction yield of carthamin in a sucrose/ChCl LMM containing 25% of water was around three times higher than in the same LMM with 10% or 50% of water. As a general trend, the extraction of polar compounds was more efficient in LMM with a high content of water than in LMM with a low content of water. In contrast, the extraction of less polar compounds can be performed in LMM with low water content. Hence, the extraction capacity of LMMs or DESs is not directly correlated to the composition and the nature of the LMMs or DESs. It also relies on the viscosity, the freezing point and the water content of the LMMs or DESs. Hence, LMMs or DESs have the ability to donate and accept protons and electrons, allowing them to strongly interact with various functional groups such as hydroxyl groups, carboxylic acids and amines. This particular property of LMMs or DESs is generally taken into account for extraction issues.

4.6   Conclusion

LMMs and DESs have the potential to provide notables advantages in green processes. Hence the atom economy of their synthesis is 100% by simply mixing two or three compounds to obtain a liquid with the formation of a hydrogen bond. These promising solvents can be used for metal-catalysed organic reactions such as redox isomerization of allylic alcohols into saturated carbonyl compounds, cycloisomerizations of unsaturated organic substrates, Diels–Alder cycloadditions and the conversion of biomass to furanic compounds. The benefits of these solvents are high yield and selectivity to the desired product. These solvents can also be used for their ability to drive the reaction to the desired products and to support high reagent loading, as shown in the conversion of biomass to HMF or furfural. Moreover, they are tolerant to water, which is an advantage for chemical reactions (i.e. biomass conversion) since purification of reagents is not required to conduct the reaction. Hence, ILs are not tolerant to water, which requires their purification before use, more specifically in the conversion of biomass. Once the products are formed in DES/LMM solvent, one challenge is their extraction. To this aim, some organic solvents are required to recover the products formed and, depending on the extraction coefficient, the recovery can be low. This is one point that has to be addressed when using DES or LMM solvents. However, one should mention that DESs/LMMs can be used to extract some products like glycerol and furanic derivatives by forming a eutectic mixture. The thermal stability and the recyclability of eutectic mixtures should be investigated since these two major issues are often neglected in the literature. The thermal stability of ChCl can be detrimental to the reaction. In the presence of a base, ChCl can be converted to trimethylamine through a Hoffman elimination reaction. Their recyclability still suffers from being an energy-consuming process, which hampers their use at an industrial scale. However, the many advantages of these new solvents are promising for the chemical industry.

References

1. Clark, J.H. (1990) Green chemistry: challenges and opportunities. *Green Chem.*, **1** (1), 1–8.
2. Anastas, P.T. and Warner, J.C. (1999) *Green Chemistry Theory and Practice*, Oxford University Press, Oxford, p. 30.
3. Tang, S.L.Y., Smith, R.L. and Poliakoff, M. (2005) Principles of green chemistry: PRODUCTIVELY. *Green Chem.*, **7** (11), 761–762.
4. Constable, D.J.C., Jiménez-González, C. and Henderson, R.K. (2007) Perspective on solvent use in the pharmaceutical industry. *Org. Process Res. Dev.*, **11** (1), 133–137.
5. Clark, J.H. and Tavener, S.J. (2007) Alternative solvents: shades of green. *Org. Process Res. Dev.*, **11** (1), 149–155.
6. Jessop, P.G. (2011) Searching for green solvents. *Green Chem.*, **13** (6), 1391–1398.
7. Laird, T. (2012) Green chemistry is good process chemistry. *Org. Process Res. Dev.*, **16** (1), 1–2.
8. Reichardt, C. (2004) *Solvents and Solvent Effects in Organic Chemistry*, Wiley-VCH, Weinheim, pp. i–xxvi.
9. Prat, D., Wells, A., Hayler, J., et al. (2016) CHEM21 selection guide of classical – and less classical – solvents. *Green Chem.*, **18** (1), 288–296.
10. Screttas, C.G. and Eastham, J.F. (1966) Solvent effects in organometallic reactions. VI. A kinetic role of base. *J. Am. Chem. Soc.*, **88** (23), 5668–5670.
11. Leitner, W. (2002) Supercritical carbon dioxide as a green reaction medium for catalysis. *Acc. Chem. Res.*, **35** (9), 746–756.
12. Han, X. and Poliakoff, M. (2012) Continuous reactions in supercritical carbon dioxide: problems, solutions and possible ways forward. *Chem. Soc. Rev.*, **41** (4), 1428–1436.
13. Welton, T. (1999) Room-temperature ionic liquids. Solvents for synthesis and catalysis. *Chem. Rev.*, **99** (8), 2071–2083.
14. Hallet, J.P. and Welton, T. (2011) Room-temperature ionic liquids: solvents for synthesis and catalysis 2. *Chem. Rev.*, **111** (5), 3508–3576.
15. Rogers, R.D. and Seddon, K.R. (2003) Ionic liquids – solvents of the future? *Science*, **302** (5646), 792–793.
16. Wasserscheid, P. and Keim, W. (2000) Ionic liquids – new 'solutions' for transition metal catalysis. *Angew. Chem., Int. Edn*, **39** (21), 3772–3789.
17. Kirchner, B. (ed.) (2010) Ionic Liquids, *Topics in Current Chemistry*, vol. 290, Springer, Berlin.
18. Seddon, K.R. (1999) Ionic liquids: designer solvents?, In *Proc. Int. George Papatheodorou Symp.*, 17–19 September, Rio, Greece, pp. 131–135.
19. Egorova, K.S. and Ananikov, V.P. (2014) Toxicity of ionic liquids: eco(cyto)activity as complicated, but unavoidable parameter for task-specific optimization. *ChemSusChem*, **7** (2), 336–360.
20. Zhang, Q., de Oliveira Vigier, K., Royer, S. and Jérôme, F. (2012) Deep eutectic solvents: syntheses, properties and applications. *Chem. Soc. Rev.*, **41** (21), 7108–7146.

21. Ruß, C. and König, B. (2012) Low melting mixtures in organic synthesis – an alternative to ionic liquids? *Green Chem.*, **14** (11), 2969–2982.

22. Abbott, A.P., Capper, G., Davies, D.L., et al. (2003) Novel solvent properties of choline chloride/urea mixtures. *Chem. Commun.*, 2003 (1), 70–71.

23. Imperato, G., Eibler, E., Niedermaier, J. and König, B. (2005) Low-melting sugar–urea–salt mixtures as solvents for Diels–Alder reactions. *Chem. Commun.*, 2005 (9), 1170–1172.

24. Seddon, K.R. (1996) Room-temperature ionic liquids: neoteric solvents for clean catalysis. *Kinet. Catal.*, **37** (5), 693–697.

25. Carriazo, D., Serrano, M.C., Gutiérrez, M.C., et al. (2012) Deep-eutectic solvents playing multiple roles in the synthesis of polymers and related materials. *Chem. Soc. Rev.*, **41** (14), 4996–5014.

26. Francisco, M., van den Bruinhorst, A. and Kroon, M.C. (2013) Low-transition-temperature mixtures (LTTMs): a new generation of designer solvents. *Angew. Chem., Int. Edn*, **52** (11), 3074–3085.

27. Smith, E.L., Abbott, A.P. and Ryder, K.S. (2014) Deep eutectic solvents (DESs) and their applications. *Chem. Rev.*, **114** (21), 11060–11082.

28. Paiva, A., Craveiro, R., Aroso, I., et al. (2014) Natural deep eutectic solvents – solvents for the 21st century. *ACS Sustain. Chem. Eng.*, **2** (5), 1063–1071.

29. Del Monte, F., Carriazo, D., Serrano, M.C., et al. (2014) Deep eutectic solvents in polymerizations: a greener alternative to conventional syntheses. *ChemSusChem*, **7** (4), 999–1009.

30. Pena-Pereira, F. and Namieśnik, J. (2014) Ionic liquids and deep eutectic mixtures: sustainable solvents for extraction processes. *ChemSusChem*, **7** (7), 1784–1800.

31. Wagle, D.V., Zhao, H. and Baker, G.A. (2014) Deep eutectic solvents: sustainable media for nanoscale and functional materials. *Acc. Chem. Res.*, **47** (8), 2299–2308.

32. Liu, P., Hao, J., Mo, L. and Zhang, Z. (2015) Recent advances in the application of deep eutectic solvents as sustainable media as well as catalysts in organic reactions. *RSC Adv.*, **5** (60), 48675–48704.

33. van Leeuwen, P.W.N.M. (ed.) (2004) *Homogeneous Catalysis: Understanding the Art*, Kluwer Academic, Dordrecht.

34. Beller, M., Renken, A. and van Santen, R.A. (eds) (2001) *Catalysis: From Principles to Applications*, Wiley-VCH, Weinheim.

35. Steinborn, D. (ed.) (2001) *Fundamentals of Organometallic Catalysis*, Wiley-VCH, Weinheim.

36. Dixneuf, P.H. and Cadierno, V. (eds) (2013) *Metal-Catalyzed Reactions in Water*, Wiley-VCH, Weinheim.

37. Dyson, P.J. and Geldbach, T.J. (2005) *Metal Catalysed Reactions in Ionic Liquids*, Springer, Dordrecht.

38. Gu, Y. and Jérôme, F. (2010) Glycerol as a sustainable solvent for green chemistry. *Green Chem.*, **12** (7), 1127–1138.

39. Díez-Álvarez, A.E., Francos, J., Lastra-Barreira, B., et al. (2011) Glycerol and derived solvents: new sustainable reaction media for organic synthesis. *Chem. Commun.*, **47** (22), 6208–6227.

40. Gu, Y. and Jérôme, F. (2010) Bio-based solvents: an emerging generation of fluids for the design of eco-efficient processes in catalysis and organic chemistry. *Chem. Soc. Rev.*, **42** (24), 9550–9570.

41. Jiang, H.-F. (2005) Transition metal-catalyzed organic reactions in supercritical carbon dioxide. *Curr. Org. Chem.*, **9** (3), 289–297.

42. Sheldon, R.A., Arends, I.W.C.E. and Henefeld, U. (2007) *Green Chemistry and Catalysis*, Wiley-VCH, Weinheim.

43. Blusztajn, J.K. (1998) Choline, a vital amine. *Science*, **281** (5378), 794–795.

44. Petkovic, M., Ferguson, J.L., Gunaratne, H.Q.N., et al. (2010) Novel biocompatible cholinium-based ionic liquids-toxicity and biodegradability. *Green Chem.*, **12** (4), 643–649.

45. Liao, J.-H., Wu, P.-C. and Bai, Y.-H. (2005) Eutectic mixture of choline chloride/urea as a green solvent in synthesis of a coordination polymer: [Zn(O_3PCH$_2$CO$_2$)]·NH$_4$. *Inorg. Chem. Commun.*, **8** (4), 390–392.

46. Zhu, A., Jiang, T., Han, B., et al. (2007) Supported choline chloride/urea as a heterogeneous catalyst for chemical fixation of carbon dioxide to cyclic carbonates. *Green Chem.*, **9** (2), 169–172.

47. Phadtare, S.B. and Shankarling, G.S. (2010) Halogenation reactions in biodegradable solvent: efficient bromination of substituted 1-aminoanthra-9,10-quinone in deep eutectic solvent (choline chloride : urea). *Green Chem.*, **12** (3), 458–462.

48. Francisco, M., van den Bruinhorst, A. and Kroon, M.C. (2012) New natural and renewable low transition temperature mixtures (LTTMs): screening as solvents for lignocellulosic biomass processing. *Green Chem.*, **14** (8), 2153–2157.

49. Zhao, H., Baker, G.A. and Holmes, S. (2011) Protease activation in glycerol-based deep eutectic solvents. *J. Mol. Catal. B: Enzym.*, **72** (3–4), 163–167.

50. Yang, D., Hou, M., Ning, H., et al. (2013) Efficient SO$_2$ absorption by renewable choline chloride–glycerol deep eutectic solvents. *Green Chem.*, **15** (8), 2261–2265.

51. Vidal, C., Suárez, F.J. and García-Álvarez, J. (2014) Deep eutectic solvents (DES) as green reaction media for the redox isomerization of allylic alcohols into carbonyl compounds catalyzed by the ruthenium complex [Ru(η^3:η^3-C$_{10}$H$_{16}$)Cl$_2$(benzimidazole)]. *Catal. Commun.*, **44**, 76–79.

52. Fürstner, A. and Davies, P.W. (2007) Catalytic carbophilic activation: catalysis by platinum and gold π acids. *Angew. Chem., Int. Edn*, **46** (19), 3410–3449.

53. Rodríguez-Álvarez, M.J., Vidal, C., Díez, J. and García-Álvarez, J. (2014) Introducing deep eutectic solvents as biorenewable media for Au(I)-catalysed cycloisomerisation of γ-alkynoic acids: an unprecedented catalytic system. *Chem. Commun.*, **50** (85), 12927–12929.

54. Vidal, C., Merz, L. and García-Álvarez, J. (2015) Deep eutectic solvents: biorenewable reaction media for Au(I)-catalysed cycloisomerisations and one-pot tandem cyclo isomerisation/Diels–Alder reactions. *Green Chem.*, **17** (7), 3870–3878.

55. Marset, X., Pérez, J.M. and Ramón, D. (2016) Cross-dehydrogenative coupling reaction using copper oxide impregnated on magnetite in deep eutectic solvents. *Green Chem.*, **18** (3), 826–833.

56. Azizi, N., Manochehri, Z., Nahayi, A. and Torkashvand, S. (2014) A facile one-pot synthesis of tetrasubstituted imidazoles catalyzed by eutectic mixture stabilized ferrofluid. *J. Mol. Liq.*, **196**, 153–158.

57. Azizi, N., Rahimi, Z. and Alipour, M. (2015) A magnetic nanoparticle catalyzed eco-friendly synthesis of cyanohydrins in a deep eutectic solvent. *RSC Adv.*, **5** (75), 61191–61198.

58. Oumahi, C., Lombard, J., Casale, S., et al. (2014) Heterogeneous catalyst preparation in ionic liquids: titania supported gold nanoparticles. *Catal. Today*, **235**, 58–71.

59. Patil, U.B., Singh, A.S. and Nagarkar, J.M. (2014) Choline chloride based eutectic solvent: an efficient and reusable solvent system for the synthesis of primary amides from aldehydes and from nitriles. *RSC Adv.*, **4** (3), 1102–1106.

60. Hajipour, A.R., Nazemzadeh, S.H. and Mohammadsaleh, F. (2014) Choline chloride/CuCl as an effective homogeneous catalyst for palladium-free Sonogashira cross-coupling reactions. *Tetrahedron Lett.*, **55** (3), 654–656.

61. Morales, R.C., Tambyrajah, V., Jenkins, P.R., et al. (2004) The regiospecific Fischer indole reaction in choline chloride·2ZnCl$_2$ with product isolation by direct sublimation from the ionic liquid. *Chem. Commun.*, 2004 (2), 158–159.

62. Azizi, N., Dezfuli, S. and Hahsemi, M.M. (2012) Eutectic salt catalyzed environmentally benign and highly efficient Biginelli reaction. *Scient. World J.*, 2012, 908702.

63. Mobinikhaledi, A. and Amiri, A. (2015) One-pot synthesis of tri- and tetrasubstituted imidazoles using eutectic salts as ionic liquid catalyst. *Res. Chem. Intermed.*, **41** (4), 2063–2070.

64. Imperato, G., Vasold, R. and König, B. (2006) Stille reactions with tetraalkylstannanes and phenyltrialkylstannanes in low melting sugar–urea–salt mixtures. *Adv. Synth. Catal.*, **348** (15), 2243–2247.

65. Imperato, G., Höger, S., Lenoir, D. and König, B. (2006) Low melting sugar–urea–salt mixtures as solvents for organic reactions – estimation of polarity and use in catalysis. *Green Chem.*, **8** (12), 1051–1055.

66. Ilgen, F. and König, B. (2009) Organic reactions in low melting mixtures based on carbohydrates and L-carnitine – a comparison. *Green Chem.*, **11** (6), 848–854.

67. Lu, J., Li, X.-T., Ma, E.-Q., et al. (2014) Superparamagnetic CuFeO$_2$ nanoparticles in deep eutectic solvent: an efficient and recyclable catalytic system for the synthesis of imidazo[1,2-*a*]pyridines. *ChemCatChem*, **6** (10), 2854–2859.

68. Jérôme, F., Ferreira, M., Bricout, H., et al. (2014) Low melting mixtures based on β-cyclodextrin derivatives and *N,N′*-dimethylurea as solvents for sustainable catalytic processes. *Green Chem.*, **16** (8), 3876–3880.

69. Ferreira, M., Jérôme, F., Bricout, H. et al. (2015) Rhodium catalyzed hydroformylation of 1-decene in low melting mixtures based on various cyclodextrins and *N,N′*-sdimethylurea. *Catal. Commun.*, **63**, 62–65.

70. Lichtenthaler, F.W. (2002) Unsaturated O- and N-heterocycles from carbohydrate feedstocks. *Acc. Chem. Res.*, **35**, 728–737.

71. Gandini, A. (1992) Polymers from renewable resources. In *Comprehensive Polymer Science*, First Supplement (eds S.L. Aggarwal and S. Russo), Pergamon Press, Oxford.

72. Kunz, M. (1993) Hydroxymethylfurfural, a possible basic chemical for industrial intermediates. In *Inulin and Inulin-Containing Crops* (ed. A. Fuchs), Elsevier, Amsterdam.

73. Gandini, A. and Belgacem, M.N. (2002) Le furfural et les polymères furaniques. *Actual. Chim.*, **261**, 56–61.

74. Moreau, C., Belgacem, M.N. and Gandini, A. (2004) Recent catalytic advances in the chemistry of substituted furans from carbohydrates and in the ensuing polymers. *Topics Catal.*, **27**, 11-30.

75. Sidhpuria, K.B., Daniel da Silva, A.L., Trindade, T. and Coutinho, J.A.P. (2011) Supported ionic liquid silica nanoparticles (SILnPs) as an efficient and recyclable heterogeneous catalyst for the dehydration of fructose to 5-hydroxymethylfurfural. *Green Chem.*, **13**, 340–349.

76. Zhang, Y., Degirmenci, V., Li, C. and Hensen, E.J.M. (2011) Phosphotungstic acid encapsulated in metal–organic framework as catalysts for carbohydrate dehydration to 5-hydroxymethylfurfural. *ChemSusChem*, **4**, 59–64.

77. Shimizu, K., Uozumi, R. and Satsuma, A. (2009) Enhanced production of hydroxymethyl-furfural from fructose with solid acid catalysts by simple water removal methods. *Catal. Commun.*, **10**, 1849–1853.

78. Tong, X. and Li, Y. (2010) Efficient and selective dehydration of fructose to 5-hydroxymethylfurfural catalyzed by Brønsted-acidic ionic liquids. *ChemSusChem*, **3**, 350–355.

79. Hansen, T.S., Mielby, J. and Riisager, A. (2011) Synergy of boric acid and added salts in the catalytic dehydration of hexoses to 5-hydroxymethylfurfural in water. *Green Chem.*, **13**, 109–114.

80. Asghari, F.S. and Yoshida, H. (2006) Acid-catalyzed production of 5-hydroxymethyl fur-fural from D-fructose in subcritical water. *Ind. Eng. Chem. Res.*, **45**, 2163–2173.

81. Asghari, F.S. and Yoshida, H. (2007) Kinetics of the decomposition of fructose catalyzed by hydrochloric acid in subcritical water: formation of 5-hydroxymethylfurfural, levulinic, and formic acids. *Ind. Eng. Chem. Res.*, **46**, 7703–7710.

82. Watanabe, M., Aizawa, Y., Iida, T. and Nishimura, R. (2005) Catalytic glucose and fructose conversions with TiO_2 and ZrO_2 in water at 473 K: relationship between reactivity and acid–base property determined by TPD measurement. *Appl. Catal. A*, **295**, 150–156.

83. McNeff, C.V., Nowlan, D.T., McNeff, L.C., et al. (2010) Continuous production of 5-hydroxymethylfurfural from simple and complex carbohydrates. *Appl. Catal. A*, **384**, 65–69.

84. Zhang, Z., Wang, Q., Xie, H., et al. (2011) catalytic conversion of carbohydrates into 5-hydroxymethylfurfural by germanium(IV) chloride in ionic liquids. *ChemSusChem*, **4**, 131–138.

85. Zhao, H., Holladay, J.E, Brown, H. and Zhang, Z.C. (2007) Metal chlorides in ionic liquid solvents convert sugars to 5-hydroxymethylfurfural. *Science*, **316**, 1597–1600.

86. Lima, S., Neves, P. and Antunes, M.M. (2009) Conversion of mono/di/polysaccharides into furan compounds using 1-alkyl-3-methylimidazolium ionic liquids. *Appl. Catal. A*, **363**, 93–99.

87. Sievers, C., Musin, I., Marzialetti, T., et al. (2009). Acid-catalyzed conversion of sugars and furfurals in an ionic-liquid phase. *ChemSusChem*, **2**, 665–671.

88. Chan, J.Y.G. and Zhang, Y. (2009) Selective conversion of fructose to 5-hydroxymethyl furfural catalyzed by tungsten salts at low temperatures. *ChemSusChem*, **2**, 731–734.

89. Qi, X., Watanabe, M., Aida, T.M. and Smith, R.L. (2009) Efficient process for conversion of fructose to 5-hydroxymethylfurfural with ionic liquids. *Green Chem.*, **11**, 1327–1331.

90. Yong, G., Zhang, Y. and Ying, J.Y. (2008) Efficient catalytic system for the selective pro-duction of 5-hydroxymethylfurfural from glucose and fructose. *Angew. Chem., Int. Edn*, **47**, 9345–9348.

91. Hu, S., Zhang, Z., Zhou, Y., et al. (2008) Conversion of fructose to 5-hydroxymethylfurfural using ionic liquids prepared from renewable materials. *Green Chem.*, **10**, 1280–1283.

92. Ilgen, F., Ott, D., Kralish, D., et al. (2009) Conversion of carbohydrates into 5-hydroxymethylfurfural in highly concentrated low melting mixtures. *Green Chem.*, **11**, 1948–1954.

93. Cao, Q., Guo, X., Guan, J., et al. (2011) A process for efficient conversion of fructose into 5-hydroxymethylfurfural in ammonium salts. *Appl. Catal. A*, **403**, 98-103.

94. Liu, F., Audemar, M., de Oliveira Vigier, K., et al. (2013) Selectivity enhancement in the aqueous acid-catalyzed conversion of glucose to 5-hydroxymethylfurfural induced by choline chloride. *Green Chem.*, **15**, 3205–3213.

95. Yang, J., de Oliveira Vigier, K., Gu, Y. and Jérôme, F. (2015). Catalytic dehydration of carbohydrates suspended in organic solvents promoted by AlCl$_3$/SiO$_2$ coated with choline chloride. *ChemSusChem.*, **3**, 269–274.

96. Mondal, D., Chaudhary, J.P., Sharma, M. and Prasad, K. (2014) Simultaneous dehydration of biomass-derived sugars to 5-hydroxymethyl furfural (HMF) and reduction of graphene oxide in ethyl lactate: one pot dual chemistry. *RSC Adv.*, **4**, 29834–29839.

97. de Oliveira Vigier, K., Benguerba, A., Barrault, J. and Jérôme, F. (2012) Conversion of fructose and inulin to 5-hydroxymethylfurfural in sustainable betaine hydrochloride-based media. *Green Chem.*, **14**, 285–289.

98. Liu, F., Barrault, J., de Oliveira Vigier, K. and Jérôme, F. (2012) Dehydration of highly concentrated solution of fructose to 5-hydroxymethylfurfural in cheap and sustainable choline chloride/CO$_2$ system. *ChemSusChem*, **5**, 1223–1226.

99. Li, X., Hou, M., Han, B., et al. (2008) Solubility of CO$_2$ in a choline chloride + urea eutectic mixture. *J. Chem. Eng. Data*, **53**, 548–550.

100. Zhao, Q., Sun, Z., Wang, S., et al. (2014) Conversion of highly concentrated fructose into 5-hydroxymethylfurfural by acid–base bifunctional HPA nanocatalysts induced by choline chloride. *RSC Adv.*, **4**, 63055–63061.

101. Hu, S., Zhang, Z., Zhou, Y., et al. (2009) Direct conversion of inulin to 5-hydroxymethyl furfural in biorenewable ionic liquids. *Green Chem.*, **11**, 873–877.

102. Assanosi, A.A., Farah, M.M., Wood, J. and Al-Duri, B. (2014) A facile acidic choline chloride–*p*-TSA DES catalysed dehydration of fructose to 5-hydroxymethylfurfural. *RSC Adv.*, **4**, 39359–39364.

103. Matsumiya, H. and Hara, T. (2015) Conversion of glucose into 5-hydroxymethylfurfural with boric acid in molten mixtures of choline salts and carboxylic acids. *Biomass Bioenergy*, **72**, 227–232.

104. da Silva Lacerda, V., López-Sotelo, J.B., Correa-Guimarães, A., et al. (2015) A kinetic study on microwave-assisted conversion of cellulose and lignocellulosic waste into hydroxymethylfurfural/furfural. *Bioresour. Technol.*, **180**, 88–96.

105. Zhang, L. and Yu, H. (2013) Conversion of xylan and xylose into furfural in biorenewable deep eutectic solvent with trivalent metal chloride added. *Bioresources*, **8**, 6014–6025.

106. Liu, F., Boissou, F., Vignault, V., et al. (2014) Conversion of wheat straw to furfural and levulinic acid in a concentrated aqueous solution of betaine hydrochloride. *RSC Adv.*, **4**, 28836–28841.

107. Kobayashi, T., Yoshino, M., Miyagawa, Y. and Adachi, S. (2015) Production of 5-hydroxymethylfurfural in a eutectic mixture of citric acid and choline chloride and its extractive recovery. *Sep. Purif. Technol.*, **155**, 26–31.

108. Krystof, M., Perez-Sanchez, M. and Dominguez de Maria, P. (2013) Lipase-catalyzed (trans)esterification of 5-hydroxymethylfurfural and separation from HMF esters using deep-eutectic solvents. *ChemSusChem.*, **6**, 630–634.

109. Qin, Y.-Z., Li, Y.-M., Zong, M.-H., et al. (2015) Enzyme-catalyzed selective oxidation of 5-hydroxymethylfurfural (HMF) and separation of HMF and 2,5-diformylfuran using deep eutectic solvents. *Green Chem.*, **17**, 3718–3722.

110. Abbott, A.P., Cullis, P.M., Gibson, M.J., et al. (2007) Extraction of glycerol from biodiesel into a eutectic based ionic liquid. *Green Chem.*, **9**, 868 –872.

111. Shahbaz, K., Mjalli, F.S., Hashim, M.A. and AlNashef, I.M. (2011) Using deep eutectic solvents based on methyl triphenyl phosphunium bromide for the removal of glycerol from palm-oil-based biodiesel. *Energy Fuels*, **25** (6), 2671–2678.
112. Shahbaz, K., Mjalli, F.S., Hashim, M.A. and AlNashef, I.M. (2011) Eutectic solvents for the removal of residual palm oil-based biodiesel catalyst. *Sep. Purif. Technol.*, **81**, 216–222.
113. Dai, Y., Witkamp, G.-J., Verpoorte, R. and Choi, Y.H. (2013) Natural deep eutectic solvents as new extraction media for phenolic metabolites in *Carthamus tinctorius* L. *Anal. Chem.*, **85**, 6272–6278.

5

Organic Carbonates: Promising Reactive Solvents for Biorefineries and Biotechnology

Paula Bracco[1] and Pablo Domínguez de María[2]

[1]*Biocatalysis, Department of Biotechnology, TU Delft, The Netherlands*

[2]*Sustainable Momentum SL, Las Palmas de Gran Canaria, Spain*

5.1 The Quest for Sustainable Solvents and the Emerging Role of Organic Carbonates

Environmental challenges, aligned with the expected depletion of fossil resources, are triggering research activities aimed at replacing today's petroleum refinery concept by more sustainable processes to deliver fuels and chemicals from renewable sources, such as lignocellulose or biological wastes. These new emerging technologies – typically gathered under the generic name of 'biorefineries' – are envisaged to focus on the full valorization of biological materials, by delivering a broad range of products, ranging from biofuels to more high-added-value materials and commodities, under mild conditions and diminished waste formation by operating in a highly integrated manner [1, 2]. For the successful implementation of these biorefineries at the practical scale, key aspects are: (i) the need for

Bio-Based Solvents, First Edition. Edited by François Jérôme and Rafael Luque.
© 2017 John Wiley & Sons Ltd. Published 2017 by John Wiley & Sons Ltd.

(bio-based) solvents, with different properties, enabling (bio)chemical reactions; (ii) the smart use (and recycling) of water on a large scale; and (iii) the introduction of (bio)catalytic steps that could provide sustainable solutions to transform feedstocks into the desired products and commodities. All in all, the proper close integration of these three aspects will define whether a given biorefinery-based system may be successful. Through that integration, it may be expected that both economic (via recycling and reuse) and environmental (reducing waste formation, or using wastes for further processing) aspects can be achieved.

In this scenario, the development and subsequent use of bio-based solvents appear to be a smart way of making use of biomass resources to produce not only the desired products, but also the solvents that will be implemented within the biorefinery. On this basis, a full petroleum-free biorefinery may be envisaged, as fossil-based solvents would be totally replaced. Following these premises, numerous examples of biogenic solvents with potential applications and useful properties have recently been reported by different research groups, with outstanding case studies ranging from biomass-derived furan-based ethers such as 2-methyltetrahydrofuran (2-MeTHF) [3], to ionic solvents such as biogenic ionic liquids and deep eutectic solvents (DESs) [4, 5]. One may well expect that these research lines will be reinforced in the coming years.

Another group of solvents that may exert tunable (reactive) properties, and that may be potentially derived from biogenic resources (e.g. natural short-chain alcohols, polyols and CO_2 or urea) are organic carbonates. In fact, depending on the substituent, a broad palette of new solvents may be envisaged, even with examples in which chirality may be introduced, thus leading to chiral solvents as well (Figure 5.1).

In this area, an outstanding example is dimethyl carbonate (DMC) [6–8]. DMC is well known as an environmentally friendly and promising methylating and methoxycarbonylating agent, enabling the replacement of hazardous methylating

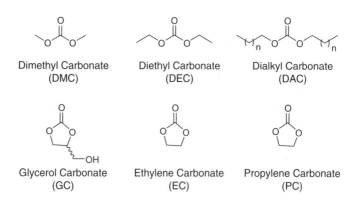

Dimethyl Carbonate
(DMC)

Diethyl Carbonate
(DEC)

Dialkyl Carbonate
(DAC)

Glycerol Carbonate
(GC)

Ethylene Carbonate
(EC)

Propylene Carbonate
(PC)

Figure 5.1 Selected examples of organic carbonates that may be used as solvents.

agents such as methyl halides, dimethyl sulfate, as well as phosgene, among many other uses [6–13]. DMC does not produce salts, and decomposes to render methanol and carbon dioxide [6–8]. Its reactivity is caused by the ambident electrophile character of DMC, displaying a reactivity that can be tuned for a variety of nucleophiles [6–8]. DMC has recently been used as solvent for different biomass-based processing activities and biocatalysis (see next sections of this chapter).

Given the potentiality that these carbonate solvents may have, their use in areas like (sustainable) catalysis has flourished, with a broad number of applications that have been extensively reviewed in the recent years [14, 15]. Herein, it must be noted that organic carbonates have been traditionally synthesized using activated carbonylic derivatives such as phosgene, or employing synthesis-gas effluents (containing CO). Given the current trends in developing novel, more sustainable routes for chemicals and materials, considerable research has been undertaken in setting up strategies that could deliver different organic carbonates at large scale and in a more sustainable manner [14–18]. In this respect, efforts have been directed to the use of raw materials such as urea [19–23], CO_2 [24–28] and glycerol [29, 30], among other relevant examples comprising fermentation-produced short-chain alcohols.

It is not the intention of this chapter to revise the applications of organic carbonates, already comprehensively reviewed elsewhere [14, 15]. Instead, selected recent applications of organic carbonates as reactive solvents in biorefineries and biotechnology will be discussed, with emphasis on the diversity of options and functions that these solvents may play: mere solvents, reagents or extractive media.

5.2 Carbonate Solvents in Biorefineries

As stated above, the possibility of using biomass-derived solvents in biorefineries is highly appealing, as fossil-based resources would then be totally replaced. Combined with efforts to introduce more sustainable syntheses for organic carbonates [14–30], it is worth investigating whether the solvent capabilities and reactivities of such solvents would fit within several biorefinery-based processes.

An area that has recently attracted interest is the valorization of lignin – through (bio)chemical modification – in which DMC can be successfully used. The rationale of the latter approach lies in the oxidation and de-aromatization of lignin using peracids [31–33], which enable the formation of modified lignin fibres with different properties for further chemical modification or uses. The size of the lignin fibres is reduced by several oxidative mechanisms, such as oxidation of hydroxyl to form carbonyl groups at the side chains, cleavage of β-aryl bonds, or modification of aromatic rings, among others [33]. Despite its potential utility, however, the storage of peracids is dangerous, and their transport and handling makes the overall process hazardous and less attractive to envisage a large-scale application at a biorefinery. To overcome this problem, *in situ* peracid formation using mineral acids or hydrolases can be considered. In these cases, peracid concentrations always remain at *catalytic* levels, and thus the associated hazards can be significantly mitigated. An outstanding example of this approach may be complete lignin oxidation and de-aromatization – starting with beech wood as raw material – to afford a lignin oil and a totally delignified polysaccharide fraction, readily accessed by cellulases. To this end, lipase B from *Candida antarctica* (CAL-B) forms *in situ* peracids starting from DMC as solvent (Scheme 5.1) [34].

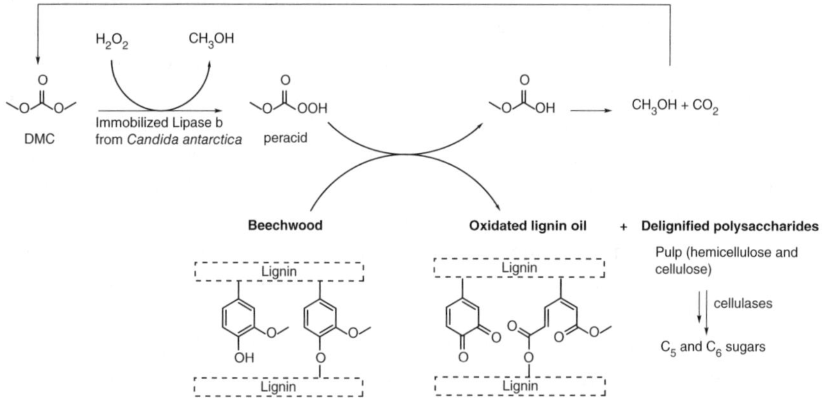

Scheme 5.1   Lipase-mediated lignin oxidation with *in situ* peracid formation in DMC (as acyl donor and solvent) [34]. Possible oxidative and de-aromatization pathways in lignin mediated by peracids are shown [33].

As observed (Scheme 5.1), the *in situ* peracid formation is achieved using diluted hydrogen peroxide and DMC as solvent as well as acyl donor for the lipase-mediated oxidation. Upon such oxidative treatment at 80°C and atmospheric pressure, a yellowish non-aromatic lignin oil (yields of ca. 50–60% compared to lignin content in beech wood) was obtained. By that approach, the production of a fully *de-aromatized* oxidized lignin *oil* rich in low-molecular-weight compounds (polyols, ketones, esters, etc.) was demonstrated for the first time. The newly discovered de-aromatized lignin derivative might be attributed to the applied non-aqueous system in DMC – in which CAL-B is fully active for synthetic purposes – in contrast to the typically conducted peracid-mediated lignin oxidation in aqueous media [31–33]. Remarkably, by opening the aromatic rings (Scheme 5.1), more flexible lignin fibres can be obtained, thus leading to an oil-like fraction. Furthermore, the remaining fully delignified white solid polysaccharide fraction obtained from the oxidative process – containing hemicellulose and cellulose – was broadly accessible to cellulases, leading to high fermentable sugar production in short reaction times (3–4 h). Therefore, in addition to the valuable lignin oil produced upon the lipase-mediated oxidation in DMC, the polysaccharide fraction might also be valuable for the isolation of fermentable sugars (C_5 and C_6) as well as for the development of new biosynthetic approaches.

Following the same oxidative approach using DMC, the Domínguez de María group also showed that operating directly with lignin (instead of with wood) leads to further advantages and options for lignin valorization via oxidation [35]. Thus, it was observed that lignin produced through the OrganoCat pre-treatment [36, 37] can be readily dissolved in DMC at high loadings (>10 g l^{-1}). Upon treatment for 30 min at 80°C under catalyst-free conditions – thus, adding diluted hydrogen peroxide to OrganoCat lignin homogeneously dissolved in DMC – different de-aromatized lignin fibres could be achieved, with the final de-aromatization value directly related to the amount of hydrogen peroxide added. Thus, several novel lignin-based fibres and oils can be produced on a simple basis on DMC [35]. In this approach, DMC was proven to play a key role in such catalyst-free lignin de-aromatization, since other solvents such as ethanol, *n*-propanol or dioxane displayed no de-aromatization under these reaction conditions. Remarkably, by using a mixture ethanol/DMC (1 : 1 v/v) the dissolution of OrganoCat lignin was increased four-fold (compared to values observed for neat DMC), and full de-aromatization occurred in 1 h upon addition of 0.2 mol H_2O_2 per gram of lignin.

Likewise, the use of DMC for lignin processing has recently found another role in the catalytic assessment of transformation and cleavage of 1,3-dilignol as model compounds for lignin. Thus, the Bolm group studied the selective transformation and cleavage of these models affording methoxybenzene and 2-aryloxyvinylbenzene derivatives in the presence of Cs_2CO_3 or LiO*t*-Bu as catalyst, respectively, employing DMC as solvent [38]. Once the optimized conditions were established, the substrate scope was investigated using more sterically hindered models. The non-catalysed reaction of dilignol (**1a**) revealed that veratrol (**4**) and **2a** were formed in trace amounts and dicarbonate **6** was isolated as the main product (Table 5.1). The Cs_2CO_3-catalysed conversion of the *threo* diastereomer **1b** of dilignol (**1a**) yielded veratrol (**4**) with 61% yield, and cleavage of *erythro* dilignol (**1c**), containing a phenolic hydroxyl group, led to the formation of **4** in slightly lower yield. Surprisingly, for monolignol (**1e**) – lacking the primary alcohol group – a significant decrease in yield was observed; presumably, steric hindrance on the arene adjacent to the benzylic hydroxyl group resulted in *cis*-alkene in good yields for both *erythro* and *threo* diastereomers **1f** and **1g**, respectively. Furthermore, when the steric hindrance was increased with **1h** and **1i**, products **7** and **8** were obtained with moderate yields, respectively. Surprisingly, the reaction conditions for **1c** and **1d** were not suitable when employing LiO*t*-Bu as catalyst; however, when using **1h** and **1i** new products were obtained with moderate yields, the *cis*-alkene (**2c**) and *cis*- and *trans*-alkene **2d**. In summary, in the Cs_2CO_3-catalysed reaction, methoxybenzene derivatives were obtained as main products in good yields, whereas a change in selectivity towards 2-aryloxyvinylbenzene derivatives was observed when an

Table 5.1 Cleavage of lignin model compounds using DMC as solvent, as recently reported by the Bolm group [38]. Reactions were carried out with 0.05 eq. of catalyst, 1.25 ml DMC at 180°C. The reaction times for catalysis by Cs$_2$CO$_3$ and LiOt-Bu were for 8 h and 12 h, respectively

Entry	Model compound	Catalyst	Product	Yield (%)
1	1a	Cs$_2$CO$_3$	**4, 5**	60 (**4**), 15 (**5**)
		LiOt-Bu	**2a, 3a**	75 (**2a**), 9 (**3a**)
2	1b	Cs$_2$CO$_3$	**4**	61
		LiOt-Bu	**4**	20
3	1c	Cs$_2$CO$_3$	**4**	45
		LiOt-Bu	No work-up possible	N.D.
4	1d	Cs$_2$CO$_3$	**4**	42
		LiOt-Bu	No work-up possible	N.D.
5	1e	Cs$_2$CO$_3$	**4**	18
		LiOt-Bu	**2a**	13
6	1f	Cs$_2$CO$_3$	**2b**	80
		LiOt-Bu	**2b**	85

Table 5.1   *(Continued)*

Entry	Model compound	Catalyst	Product	Yield (%)
7	**1g** (OMe, OH, OMe, HO, MeO, OMe)	Cs_2CO_3	**2b**	82
		LiOt-Bu		81
8	**1h** (MeO, OH, OMe, HO, OMe, OMe)	Cs_2CO_3	**7** (OMe, MeO, OMe)	57
		LiOt-Bu	**2c** (MeO, MeO, O, OMe, OMe)	82
9	**1i** (OMe, OH, OMe, OMe, OH, OMe, OMe)	Cs_2CO_3	**8** (OMe, OMe, OMe)	62
		LiOt-Bu	**2d** (OMe, O, OMe, OMe, OMe, OMe)	44 (*cis*), 20 (*trans*)

increase in steric hindrance at the aryl ring adjacent to the benzylic hydroxyl group occurred. Furthermore, the same selectivity was observed in LiOt-Bu-catalysed reactions using non-phenolic ring models.

In addition to the above-reported lignin valorization approaches using carbonates as (reactive) solvents, recently it was demonstrated that organic carbonates can also be successfully applied as solvents for the pre-treatment of lignocellulosic biomass to improve saccharification procedures for the production of fermentable sugars. Thus, Zhang *et al.* studied thoroughly the use of solvent systems based on acidified mixtures of cyclic alkyl carbonates – such as glycerol carbonate (GC) and ethylene carbonate (EC), among others – and glycerol to treat sugarcane bagasse [39]. The glucan digestibility (per cent) and total glucose yield obtained (per cent) were indicators of the impact of the treatment in delignification and defibrillation. Results obtained by pre-treatment of sugarcane bagasse at 90°C for 30 min with acidified solvents (1.2% H_2SO_4) followed by enzymatic hydrolysis showed that both EC and GC pre-treatments led to higher glucan digestibility as compared to treatments with mixed carbonate and glycol solvents. Moreover, GC showed 90% of glucan digestibility and 80 % glucose yield, whereas only 16% and 15% were obtained, respectively, with acidified EC treatment. These results provide significantly better performance when GC solvent is used. The difference of pre-treatment effectiveness between both tested carbonate solvents (GC and EC) can most likely be attributed to the OH functionality present in the organic solvent GC, enhancing in this way its capacity to delignify and defibrillate biomass. Microcrystalline cellulose (MCC) was also treated under the same conditions followed by application of cellulases. Both systems revealed cellulose recoveries in the range of 87–93%, close to glucan recoveries of pre-treated bagasse. This suggests that alkylene carbonate/alkylene glycol systems do not hydrolyse glucan significantly under tested conditions. The highest amounts of glucose were obtained with alkylene carbonates as pure solvents, indicating their positive effect on cellulose depolymerization.

Apart from pre-treatment of lignocellulosic materials, organic carbonates are also finding use in biorefinery-type processes such as the transformation of biogenic acids like levulinic or succinic acids [1, 2]. Herein, levulinic acid (LA) is considered an important environmentally bio-based C$_5$ feedstock due to its broad availability and renewable origin. LA can be efficiently synthesized by several routes starting from cellulose, for example in aqueous reaction media via a Brønsted acid-catalysed reaction sequence [1]. Several potentially marketable products, namely 2-methyltetrahydrofuran, γ-valerolactone, 1,4-pentadiol, diphenolic acid, δ-aminolevulinic aid, ethyl levulinate and succinic acid, among others, are produced via homogeneous and heterogeneous catalysis using LA as starting material [1]. With regard to organic carbonates, the Perosa group has reported a new DMC-based chemical process to transform LA under basic conditions into potential chemical building blocks such as methyl levulinate and dimethyl succinate, using DMC as reactive solvent [40]. Full conversion of LA was obtained using catalytic amounts of K$_2$CO$_3$ (LA/K$_2$CO$_3$ of 1 : 2), a molar ratio of LA/DMC of 1 : 20, at 200°C for 6 h. Unfortunately, the results showed a high formation of unknown products (73%) and only low conversion into the desired product (21%) (Scheme 5.2). Surprisingly, by reducing the reaction time and temperature to 4 h and 160°C, respectively, dimethyl succinate was successfully produced with 99% conversion. Furthermore, the addition of methanol (molar ratio MeOH/DMC 1 : 1) led to the unexpected formation of the dimethyl ketal of methyl levulinate with a yield of 20%. In conclusion, by using different reaction conditions (time, temperature, molar ratios, co-solvents) as well as the use of different catalysts, the reaction could be tuned towards the rather selective formation of different products.

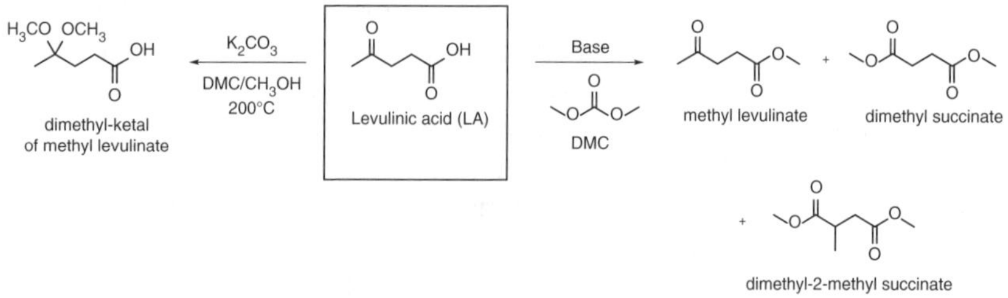

Scheme 5.2 Production of methyl levulinate and dimethyl succinate via a DMC-based catalysed process using LA derived from biomass [40].

Likewise, the Straathof group have focused on the entire bio-based succinate production process, integrated with its recovery by ion exchange, and further upgrade to dimethyl succinate (DMS) by direct downstream catalysis was established [41]. Owing to its chemical functionality, succinate esters (particularly DMS) are important precursors for several petrochemical products and can also be applied as monomers in polymer chemistry. In this process the succinate anion – produced by fermentation – is captured from the fermentative aqueous solution by a strong anion exchange resin in (bi)carbonate form, releasing the respective salt, which can be consequently used as a carbon source and a neutralization agent in the fermentation step. Subsequently, adsorbed succinate is alkylated at both carboxylate groups by a quaternary ammonium-based catalysis (Q^+) to which succinate anion is electrostatically bound, using DMC as reagent as well as the solvent. Thus DMS is formed and the anion exchange material is kept in a proper ionic form in order to retain electroneutrality. Kinetic determinations showed yields of 0.92 mol DMS produced per mole of succinate after 5 h at 120°C, described by a pseudo-first-order model. Although lower temperatures significantly reduced the reaction rates, DMS was still formed at 80°C (below the DMC boiling point) with a yield of 0.2 mol DMS per mole of succinate after 8 h of reaction. A two-step reaction mechanism was proposed where the diester is released from the resin only after methylation. The main advantage of the integrated process using DMC as solvent is that it avoids the production of stoichiometric divalent succinate salts; therefore chemical consumption and waste production are significantly reduced. Moreover, by-products derived from the decomposition of DMC can be recycled, maximizing the carbon usage.

In another area, aimed at establishing synthetic pathways to upgrade phenolic fractions derived from lignin processing, Perosa *et al.* have also investigated the methylation and transesterification of *p*-coumaryl model alcohols, such as cinnamyl alcohol (**9**) and 4-(3-hydroxypropyl)phenol (**10**), in a base-catalysed reaction using DMC as reagent/solvent [42]. The selectivity of methylation/decarboxylation on the aliphatic and phenolic OH sites of these model alcohols in the presence of DMC could be tuned by modifying the catalyst and/or reaction temperatures. In detail, K_2CO_3, $CsF/\alpha Al_2O_3$, NaX, NaY and ionic liquid [P_{8881}][CH_3OCOO] were tested as catalysts under different reaction times (up to 240 h) and temperatures (70–180°C), using DMC as acyl donor as well as solvent. Base-catalysed reactions using K_2CO_3, $CsF/\alpha Al_2O_3$ and [P_{8881}][CH_3OCOO] yielded aliphatic carbonates (**9a** and **10c**, Scheme 5.3) at 90°C by transesterification with high conversion and selectivity, more than 90% and 80%, respectively. Furthermore, the respective methyl ethers (**9b**, 90% conversion, 91% selectivity) were instead produced at higher temperatures (180°C) when applying faujasite NaY as catalyst. Nevertheless, different activities towards cinnamyl alcohol and 4-(3-hydroxypropyl)phenol were observed when using the more basic NaX. In the first case, the reaction occurred at the carbonyl carbon of DMC, forming the methyl carbonate intermediate **9a**, followed by *in situ* decarboxylation to the methyl ether **9b**, in contrast to the NaY catalyst, which produced **9b** directly. Similarly, NaX displayed higher yields for 4-(3-hydroxypropyl) phenol than NaY catalyst. However, the products **10e** and **10d** were obtained through the direct methylation of both the aliphatic and aromatic hydroxyls and via transesterification of the aliphatic hydroxyl and aromatic OH methylation with DMC, respectively (Scheme 5.3). Thus, lignin-like model compounds, such as cinnamyl alcohol and 4-(3-hydroxypropyl)phenol, were successfully upgraded by applying a base-catalysed approach where DMC acts as acyl donor as well as solvent in the reaction.

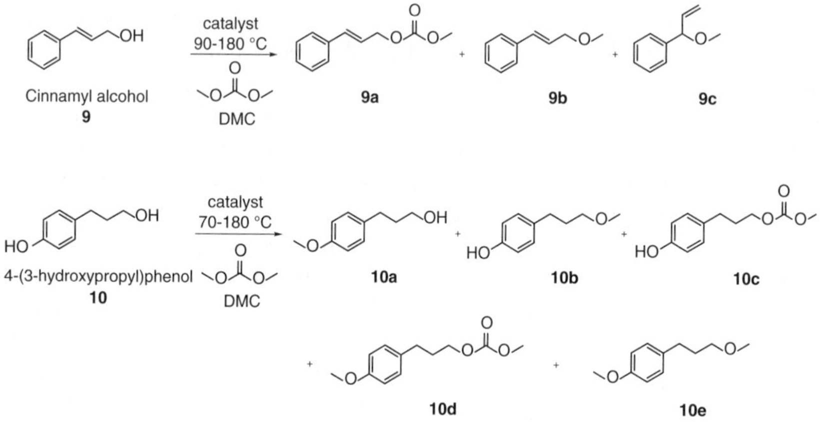

Scheme 5.3   Base-catalysed methylation/transesterification of cinnamyl alcohol (**9**) and 4-(3-hydroxypropyl)phenol (**10**) using DMC as reagent/solvent [42].

5.3   Biotechnology: from Enzymatic Synthesis of Organic Carbonates to Enzyme Catalysis in these Non-Conventional Media

As stated above, organic carbonates exert a potential double profile: on the one hand, they may be employed as solvents; on the other, their intrinsic reactivity may be used to trigger processes (e.g. acting as reagents while dissolving other substrates). Within biocatalysis, this has been the case of DMC, which has attracted considerable interest as solvent and as reagent to produce, for instance, glycerol carbonate. Thus, while a potentially useful waste like glycerol is valorized [43, 44], novel organic carbonates are produced.

In this area, several research groups [30, 45–47] have reported on using lipases (mostly CAL-B) to perform transesterification of DMC with glycerol to produce glycerol carbonate. In one of the examples [45], DMC was proposed to be used as both solvent and reagent, thus providing more sustainable (and industrially sound) conditions for the production of glycerol carbonate. By means of immobilized

CAL-B, more than 90% conversion with >90% selectivity were obtained in 48 h under optimized mild conditions (70°C and atmospheric pressure) (Scheme 5.4). Given the immiscibility of glycerol in DMC, a negative mixing effect in the overall conversion caused by the excess amount of DMC was observed. Thus, a molar ratio of 10 (DMC over glycerol adsorbed in silica) was found to give the best conditions for the solvent-free transesterification. Furthermore, the proven inhibitory effect of methanol in the synthesis rate was overcome by its *in situ* removal with molecular sieves. Moreover, Tudorache *et al.* also focused their research on finding and optimizing the use of more lipases (extracted from 12 different microorganisms) able to perform the synthesis of glycerol carbonate in solvent-free reactions [47]. Consistent with previous process conditions, it was confirmed that a 10-fold excess of DMC (compared to glycerol) leads to higher conversions and selectivities, in this case using *Aspergillus niger* lipase, which proved to be a better catalyst than CAL-B for that process. Other authors have considered the use of bases, such as K_2CO_3, for the same synthesis of glycerol carbonate in solvent-free (DMC) [48] (Scheme 5.4).

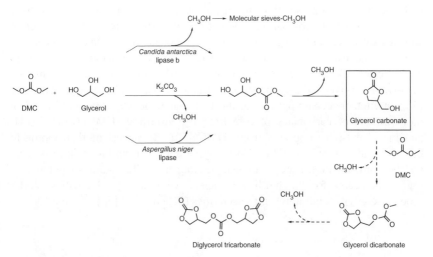

Scheme 5.4 Synthesis of carbonate solvents using DMC as reaction medium and CAL-B [46], *Aspergillus niger* lipase [47] or K_2CO_3 [48] as catalysts. Dashed lines depict the observed formation of secondary products.

The reported examples demonstrate the impact on conversion and selectivity when using different catalysts – inorganic structure or enzyme – and reaction conditions, given the ambient role of DMC as solvent/reagent. Furthermore, the fact that DMC acts also as acyl donor should not be neglected. Particularly in the lipase-catalysed syntheses, the size of the alkyl group of the carbonate molecule will have a direct impact on the lipase's affinity towards the acyl donor since they are naturally designed to accommodate hydrophobic fatty acid chains. Thus, Cushing and Peretti demonstrated that, when using dimethyl, diethyl and dibutyl carbonate as substrates in the CAL-B-catalysed synthesis of glycerol carbonate using *tert*-butanol as solvent, conversions after 48 h were 13%, 48% and 66%, respectively [43].

Other studies in the area have focused on process optimization to tackle the economics of the overall reaction. Thus, the immobilization and recyclability of lipase nanoparticles was assessed for the synthesis of glycerol carbonate in DMC [49]. Likewise, the enzymatic co-production of biodiesel and glycerol carbonate by using DMC and soybean oil as substrates was successfully shown as well [50]. More recently, the process has been combined with other process set-up conditions, such as ultrasound, to enhance the yields and conversions of the process [51], or microwaves for the formation of analogous propylene carbonate (PC) starting from DMC and 1,2-propanediol [52].

From the examples described above, it can be envisaged that organic carbonate can perform the roles of solvent, extractive agent or reagent depending on the desired application. Apart from producing other carbonates (e.g. glycerol carbonates), several applications in organic synthesis (mostly with lipases) have been studied as well, comparing different esterification reactions using different solvents (also water) with several organic carbonates [53]. Other applications comprise the use of fatty acids as substrates for the formation of fatty acid methyl esters [54], or the assessment of immobilized enzymes in the synthesis of 3-ethyl-1,3-oxazolidin-2-one using 2-amino-alcohol and dimethyl carbonate [55]. In another recent application, the Domínguez de María group has shown the synthesis of carbonates of 5-hydroxymethylfurfural (HMF) with CAL-B and in solvent-free processes using DMC [56]. By means of that approach, the two fields, biorefineries (as HMF may be derived from biogenic resources) and biotransformations, are combined, leading to the formation of potentially useful molecules for the bio-based society, with high productivities (under proof-of-concept conditions) and selectivities (Scheme 5.5) [56].

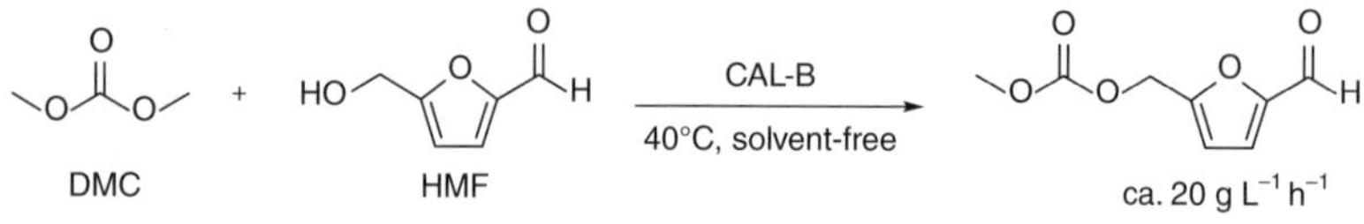

Scheme 5.5 Lipase-catalysed transesterification of DMC with HMF to afford HMF-based carbonates [56].

In the previous section on biorefineries it was shown that lipases can accept DMC as substrate to form peroxycarboxylic acids in the presence of diluted hydrogen peroxide, which may be used for lignocellulose oxidation [34]. The same lipase-mediated approach has been shown to be useful for the Prilezhaev epoxidation of alkenes to afford different epoxides, as well as for Baeyer–Villiger reactions under extremely mild reaction conditions [57].

Apart from hydrolases (mostly lipases), other enzymes like glycosidases have also found applications in organic carbonates as solvents [58]. As a relevant example, the regioselectivity in the transglycosilation of *N*-acetylglucosamine with *p*-nitrophenyl-β-D-galactopyranoside catalysed by Biolact No. 5 β-galactosidase was found to be highly affected by the type and amount of co-solvent added to the reaction. Interestingly, the use of glycerol-based solvents modified the traditional regioselectivity of this biocatalyst in favour of the β-(1→4) to β-(1→6) linker product. In particular, adding 0.5 M of 2-hydroxytrimethylene carbonate to the 50 mM citrate/phosphate buffer yielded 75% β-(1→4); however, at higher concentrations (1 M) the regioselectivity changed towards the β-(1→6) product with 45% yield. At higher co-solvent concentrations (up to 5 M), hydrolysis of the donor molecule occurred yielding 81% of galactose [59].

5.4 Concluding Remarks

Organic carbonates hold promising prognoses for their future use in sustainable chemistry. Owing to their structures, they may serve as mere solvents, as reagents, or even as extractive media for different processes. In this chapter, relevant recent examples covering their use in biorefineries and in biotransformations have been discussed. By smartly applying the carbonate properties, many effects and advantages can be achieved. As points for future consideration and improvement, the need for setting up environmentally friendly syntheses for these carbonates appear to be crucial, if applications at large scale are to be envisaged. In that direction, considerable efforts are being made to produce organic carbonates in a sustainable manner, to be subsequently applied for further (bio)chemical transformation with high ecological standards.

References

1. Kamm, B., Gruber, P.R. and Kamm, M. (2000) *Biorefineries – industrial processes and products.* In *Ullmann's Encyclopedia of Industrial Chemistry*, Wiley-VCH, Weinheim.
2. Tuck, C.O., Pérez, E., Horváth, I.T., *et al.* (2012) Valorization of biomass: deriving more value from waste. *Science*, **337** (6095), 695–699.
3. Pace, V., Hoyos, P., Castoldi, L., *et al.* (2012) 2-Methyltetrahydrofuran (2-MeTHF): a biomass-derived solvent with broad application in organic chemistry. *ChemSusChem*, **5** (8), 1369–1379.
4. Gu, Y. and Jérôme, F. (2013) Bio-based solvents: an emerging generation of fluids for the design of eco-efficient processes in catalysis and organic chemistry. *Chem. Soc. Rev.*, **42** (24), 9550–9570.
5. Domínguez de María, P. (2014) Recent trends in (ligno)cellulose dissolution using neoteric solvents: switchable, distillable and bio-based ionic liquids. *J. Chem. Technol. Biotechnol.*, **89** (1), 11–18.

6. Aricò, F. and Tundo, P. (2010) Dimethyl carbonate as a modern green reagent and solvent. *Russ. Chem. Rev.*, **79** (6), 479.

7. Tundo, P., Rossi, L. and Loris, A. (2005) Dimethyl carbonate as an ambident electrophile. *J. Org. Chem.*, **70** (6), 2219–2224.

8. Pacheco, M.A. and Marshall, C.L. (1997) Review of dimethyl carbonate (DMC) manufacture and its characteristics as a fuel additive. *Energy Fuels*, **11** (1), 2–29.

9. Bomben, A., Selva, M., Tundo, P. and Valli, L. (1999) A continuous-flow O-methylation of phenols with dimethyl carbonate in a continuously fed stirred tank reactor. *Ind. Eng. Chem. Res.*, **38** (5), 2075–2079.

10. Selva, M., Militello, E. and Fabris, M. (2008) The methylation of benzyl-type alcohols with dimethyl carbonate in the presence of Y- and X-faujasites: selective synthesis of methyl ethers. *Green Chem.*, **10** (1), 73–79.

11. Fabris, M., Lucchini, V., Noè, M., *et al.* (2009) Ionic liquids made with dimethyl carbonate: solvents as well as boosted basic catalysts for the Michael reaction. *Chemistry, Eur. J.*, **15** (45), 12273–12282.

12. Selva, M., Fabris, M. and Perosa, A. (2011) Decarboxylation of dialkyl carbonates to dialkyl ethers over alkali metal-exchanged faujasites. *Green Chem.*, **13** (4), 863.

13. Selva, M., Benedet, V. and Fabris, M. (2012) Selective catalytic etherification of glycerol formal and solketal with dialkyl carbonates and K_2CO_3. *Green Chem.*, **14** (1), 188–200.

14. Schäffner, B., Schäffner, F., Verevkin, S.P. and Börner, A. (2010) Organic carbonates as solvents in synthesis and catalysis. *Chem. Rev.*, **110** (8), 4554–4581.

15. Shaikh, A.-A.G. and Sivaram, S. (1996) Organic carbonates. *Chem. Rev.*, **96** (3), 951–976.

16. Aresta, M. and Galatola, M. (1999) Life cycle analysis applied to the assessment of the environmental impact of alternative synthetic processes. The dimethylcarbonate case: part 1. *J. Cleaner Prod.*, **7** (3), 181–193.

17. Huang, S., Yan, B., Wang, S. and Ma, X. (2015) Recent advances in dialkyl carbonates synthesis and applications. *Chem. Soc. Rev.*, **44** (10), 3079–3116.

18. Martín, C., Fiorani, G. and Kleij, A.W. (2015) Recent advances in the catalytic preparation of cyclic organic carbonates. *ACS Catal.*, **5** (2), 1353–1370.

19. Zhao, X., Jia, Z. and Wang, Y. (2006) Clean synthesis of propylene carbonate from urea and 1,2-propylene glycol over zinc–iron double oxide catalyst. *J. Chem. Technol. Biotechnol.*, **81** (5), 794–798.

20. Zhao, X., Zhang, Y. and Wang, Y. (2004) Synthesis of propylene carbonate from urea and 1,2-propylene glycol over a zinc acetate catalyst. *Ind. Eng. Chem. Res.*, **43** (15), 4038–4042.

21. Bhanage, B.M., Fujita, S., Ikushima, Y. and Arai, M. (2003) Transesterification of urea and ethylene glycol to ethylene carbonate as an important step for urea based dimethyl carbonate synthesis. *Green Chem.*, **5** (4), 429.

22. Wang, M., Zhao, N., Wei, W. and Sun, Y. (2004) Synthesis of dimethyl carbonate from urea and methanol over metal oxides. In *Studies in Surface Science and Catalysis*, vol. 153, Elsevier, Amsterdam, pp. 197–200.

23. Wang, M., Zhao, N., Wei and Sun, Y. (2005) Synthesis of dimethyl carbonate from urea and methanol over ZnO. *Ind. Eng. Chem. Res.*, **44** (19), 7596–7599.

24. Sakakura, T., Choi, J.-C. and Yasuda, H. (2007) Transformation of carbon dioxide. *Chem. Rev.*, **107** (6), 2365–2387.

25. Sakakura, T., Choi, J.-C., Saito, Y., *et al.* (1999) Metal-catalyzed dimethyl carbonate synthesis from carbon dioxide and acetals. *J. Org. Chem.*, **64** (12), 4506–4508.

26. Kohno, K., Choi, J.-C., Ohshima, Y., *et al.* (2008) Reaction of dibutyltin oxide with methanol under CO_2 pressure relevant to catalytic dimethyl carbonate synthesis. *J. Organomet. Chem.*, **693** (7), 1389–1392.

27. Aresta, M. and Dibenedetto, A. (2007) Utilisation of CO_2 as a chemical feedstock: opportunities and challenges. *Dalton Trans.*, 2007 (28), 2975–2992.

28. Sakakura, T. and Kohno, K. (2009) The synthesis of organic carbonates from carbon dioxide. *Chem. Commun.*, 2009 (11), 1312–1330.

29. Behr, A., Eilting, J., Irawadi, K., *et al.* (2008) Improved utilisation of renewable resources: new important derivatives of glycerol. *Green Chem.*, **10** (1), 13–30.

30. Kim, S.C., Kim, Y.H., Lee, H., *et al.* (2007) Lipase-catalyzed synthesis of glycerol carbonate from renewable glycerol and dimethyl carbonate through transesterification. *J. Mol. Catal. B: Enzym.*, **49** (1–4), 75–78.

31. Yin, D., Jing, Q., AlDajani, W.W., *et al.* (2011) Improved pretreatment of lignocellulosic biomass using enzymatically-generated peracetic acid. *Bioresour. Technol.*, **102** (8), 5183–5192.

32. Duncan, S., Jing, Q., Katona, A., *et al.* (2009) Increased saccharification yields from aspen biomass upon treatment with enzymatically generated peracetic acid. *Appl. Biochem. Biotechnol.*, **160** (6), 1637–1652.

33. Brasileiro, L.B., Colodette, J.L. and Piló-Veloso, D. (2001) The use of peracids in delignification and cellulose pulp bleaching. *Química Nova*, **24** (6), 819–829.

34. Wiermans, L., Pérez-Sánchez, M. and Domínguez de María, P. (2013) Lipase-mediated oxidative delignification in non-aqueous media: formation of de-aromatized lignin-oil and cellulase-accessible polysaccharides. *ChemSusChem*, **6** (2), 251–255.

35. Wiermans, L., Schumacher, H., Klaaßen, C.-M. and Domínguez de María, P. (2015) Unprecedented catalyst-free lignin dearomatization with hydrogen peroxide and dimethyl carbonate. *RSC Adv.*, **5** (6), 4009–4018.

36. Vom Stein, T., Grande, P.M., Kayser, H., *et al.* (2011) From biomass to feedstock: one-step fractionation of lignocellulose components by the selective organic acid-catalyzed depolymerization of hemicellulose in a biphasic system. *Green Chem.*, **13** (7), 1772.

37. Grande, P.M., Viell, J., Theyssen, N., *et al.* (2015) Fractionation of lignocellulosic biomass using the OrganoCat process. *Green Chem.*, **17** (6), 3533–3539.

38. Dabral, S., Mottweiler, J., Rinesch, T. and Bolm, C. (2015) Base-catalysed cleavage of lignin β-O-4 model compounds in dimethyl carbonate. *Green Chem.*, **17** (11), 4908–4912.

39. Zhang, Z., Rackemann, D.W., Doherty, W.O.S. and O'Hara, I.M. (2013) Glycerol carbonate as green solvent for pretreatment of sugarcane bagasse. *Biotechnol. Biofuels*, **6** (1), 153.

40. Caretto, A. and Perosa, A. (2013) Upgrading of levulinic acid with dimethylcarbonate as solvent/reagent. *ACS Sustain. Chem. Eng.*, **1** (8), 989–994.

41. López-Garzón, C.S., van der Wielen, L.A.M. and Straathof, A.J.J. (2014) Green upgrading of succinate using dimethyl carbonate for a better integration with fermentative production. *Chem. Eng. J.*, **235**, 52–60.

42. Stanley, J.N.G., Selva, M., Masters, A.F., *et al.* (2013) Reactions of *p*-coumaryl alcohol model compounds with dimethyl carbonate. Towards the upgrading of lignin building blocks. *Green Chem.*, **15** (11), 3195.

43. Cushing, K.A. and Peretti, S.W. (2013) Enzymatic processing of renewable glycerol into value-added glycerol carbonate. *RSC Adv.*, **3** (40), 18596.

44. Pagliaro, M., Ciriminna, R., Kimura, H., *et al.* (2007) From glycerol to value-added products. *Angew. Chem., Int. Edn*, **46** (24), 4434–4440.

45. Lee, K.H., Park, C.-H. and Lee, E.Y. (2010) Biosynthesis of glycerol carbonate from glycerol by lipase in dimethyl carbonate as the solvent. *Bioprocess. Biosyst. Eng.*, **33** (9), 1059–1065.

46. Ochoa-Gómez, J.R., Gómez-Jiménez-Aberasturi, O., Maestro-Madurga, B., *et al.* (2009) Synthesis of glycerol carbonate from glycerol and dimethyl carbonate by transesterification: catalyst screening and reaction optimization. *Appl. Catal. A: Gen.*, **366** (2), 315–324.

47. Tudorache, M., Protesescu, L., Coman, S. and Parvulescu, V.I. (2012) Efficient bio-conversion of glycerol to glycerol carbonate catalyzed by lipase extracted from *Aspergillus niger. Green Chem.*, **14** (2), 478.

48. Rokicki, G., Rakoczy, P., Parzuchowski, P. and Sobiecki, M. (2005) Hyperbranched aliphatic polyethers obtained from environmentally benign monomer: glycerol carbonate. *Green Chem.*, **7** (7), 529.

49. Tudorache, M., Protesescu, L., Negoi, A. and Parvulescu, V.I. (2012) Recyclable biocatalytic composites of lipase-linked magnetic macro-/nano-particles for glycerol carbonate synthesis. *Appl. Catal. A: Gen.*, **437–438**, 90–95.

50. Seong, P.-J., Jeon, B.W., Lee, M., *et al.* (2011) Enzymatic coproduction of biodiesel and glycerol carbonate from soybean oil and dimethyl carbonate. *Enzyme Microbial Technol.*, **48** (6–7), 505–509.

51. Waghmare, G.V., Vetal, M.D. and Rathod, V.K. (2015) Ultrasound assisted enzyme catalyzed synthesis of glycerol carbonate from glycerol and dimethyl carbonate. *Ultrasonics Sonochem.*, **22**, 311–316.

52. Yadav, G.D., Hude, M.P. and Talpade, A.D. (2015) Microwave assisted process intensification of lipase catalyzed transesterification of 1,2-propanediol with dimethyl carbonate for the green synthesis of propylene carbonate: novelties of kinetics and mechanism of consecutive reactions. *Chem. Eng. J.*, **281**, 199–208.

53. Ou, G., He, B. and Yuan, Y. (2011) Lipases are soluble and active in glycerol carbonate as a novel biosolvent. *Enzyme Microbial Technol.*, **49** (2), 167–170.

54. Su, E.-Z., Zhang, M.-J., Zhang, J.-G., *et al.* (2007) Lipase-catalyzed irreversible transesterification of vegetable oils for fatty acid methyl esters production with dimethyl carbonate as the acyl acceptor. *Biochem. Eng. J.*, **36** (2), 167–173.

55. Yadav, G.D. and Pawar, S.V. (2014) Novelty of immobilized enzymatic synthesis of 3-ethyl-1,3-oxazolidin-2-one using 2-aminoalcohol and dimethyl carbonate: mechanism and kinetic modeling of consecutive reactions. *J. Mol. Catal. B: Enzym.*, **109**, 62–69.

56. Krystof, M., Pérez-Sánchez, M. and Domínguez de María, P. (2013) Lipase-catalyzed (trans)esterification of 5-hydroxy-methylfurfural and separation from HMF esters using deep-eutectic solvents. *ChemSusChem*, **6** (4), 630–634.

57. Rüsch gen. Klaas, M. and Warwel, S. (1999) Chemoenzymatic epoxidation of alkenes by dimethyl carbonate and hydrogen peroxide. *Org. Lett.*, **1** (7), 1025–1026.

58. Pérez-Sánchez, M., Sandoval, M., Hernáiz, M.J. and Domínguez de María, P. (2013) Biocatalysis in biomass-derived solvents: the quest for fully sustainable chemical processes. *Curr. Org. Chem.*, **17** (11), 1188–1199.

59. Pérez-Sánchez, M., Sandoval, M. and Hernáiz, M.J. (2012) Bio-solvents change regioselectivity in the synthesis of disaccharides using Biolacta β-galactosidase. *Tetrahedron*, **68** (9), 2141–2145.

6

Life Cycle Assessment for Green Solvents

Philippe Loubet, Michael Tsang, Eskinder Gemechu, Amandine Foulet, and Guido Sonnemann

Institut des Sciences Moléculaires, Université de Bordeaux, Talence, France

6.1 Introduction

Solvents are an important aspect of many industries, including chemicals, paints, coatings, pharmaceuticals and agriculture. Globally, they are used in large quantities each year, with more than US$25 billion in solvent sales in 2013 [1]. Because they are used in such large quantities, they have the potential to contribute significantly to the environmental impacts of products and processes. Since the introduction of the green chemistry concept [2], new solvents have been developed with the general aim of improving their environmental performance. Current efforts used in the design of solvents that meet the green chemistry criteria are included in Table 6.1 [3].

Green solvents aim to replace traditionally harmful and environmentally unfriendly solvents, such as those that are toxic/carcinogenic (e.g. benzene) or that contribute to ozone depletion (e.g. chlorofluorocarbons, CFCs) and photochemical smog formation (e.g. volatile organic compounds, VOCs), for example [4]. So far, much of the discussion on green solvents has revolved around only a single aspect of the principles of green chemistry, such as use of renewable feedstocks, for example, which, although they are sourced from non-fossil fuel

Bio-Based Solvents, First Edition. Edited by François Jérôme and Rafael Luque.
© 2017 John Wiley & Sons Ltd. Published 2017 by John Wiley & Sons Ltd.

Table 6.1 *Brief overview of green chemistry criteria and options, and their correlation to the 12 principles of green chemistry (based on Capello et al. [3]).*

Safer solvents options	Description	Principles of green chemistry
Substitution of hazardous solvents for safer ones	Solvents can be hazardous to health and safety, but the substitution of hazardous solvents with safer ones such as water can avoid one or more of these issues	Benign solvents Safer solvents and auxiliaries Designing for safer chemicals Less hazardous chemical synthesis Prevention
Use of bio-based solvents whose feedstocks are renewable	Many solvents are petroleum-based products that could be substituted with ones from bio-based feedstocks, such as the production of acetates and carbonates as the by-product of bio-based ethanol production	Use of renewable feedstocks
Supercritical fluids	Supercritical solvents such as CO_2 involve normal solvents being raised to their supercritical temperatures and pressures and are notable for being easy to separate after reaction and increase reaction rates	Designing for energy efficiency Prevention
Fluorous solvents	Fluorous solvents such as perfluorocarbons are unique in their partitioning abilities and have high selectivity and are fairly unreactive	Safer solvents and auxiliaries Inherent chemical safety for accident prevention Prevention
Ionic liquids	These are salts that are in the liquid state at moderate or room temperatures and are known to have low volatility compared to other conventional solvents	Designing for energy efficiency Prevention
No solvent	N.A.	Safer solvents and auxiliaries Less hazardous chemical synthesis Designing for energy efficiency Inherent chemical safety for accident prevention

sources, may ultimately lead to higher greenhouse gas emissions from upstream production and processing of the renewable feedstock. Furthermore, a solvent might reduce its overall greenhouse gas emissions but could be more toxic to workers handling the substance. Therefore, it is necessary to look at presumed 'green' solvents from a more holistic viewpoint involving life cycle thinking (Figure 6.1) and multiple environmental criteria in order to avoid burden shifting.

In this context, there is a need for the consistent application of tools to assess the environmental performance of green solvents. Previous tools have been developed

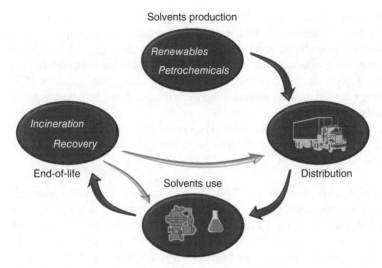

Figure 6.1 The life cycle of a solvent. (A colour version of this figure appears in the plate section.)

for conventional solvents, such as the solvent selection guide [5], which helps chemists with the consideration of health and safety criteria. An updated solvent selection guide was developed to broaden the perspective and the scope of the guide using environmental criteria [6]. However, for a rigorous, transparent and more powerful quantitative approach, an environmental assessment tool such as life cycle assessment (LCA) should be considered for all potential green solvents. LCA is a well-vetted and widely used environmental management tool [7] that can assist the evaluation and reduction of environmental impacts from industrial activities across its entire life cycle from materials sourcing to end-of-life issues (Figure 6.1).

The objectives of this chapter are to introduce the reader to LCA as a methodology and to demonstrate its applicability to both conventional and potentially green solvents. The associated structure of this chapter is as follows:

- to introduce and explain the LCA methodology (section 6.2),
- to provide a critical review of the existing literature on LCA applied to conventional and green solvents (sections 6.3 and 6.4), and
- to discuss the lessons learnt from the review and the main challenges to increase the relevance of the environmental assessment of green solvents (section 6.5).

6.2   Life Cycle Assessment: An Overview

LCA is a tool used to quantify a wide range of environmental and human health impacts of a good or service across all life cycle stages (i.e. cradle to grave), including raw material extraction, materials processing, product manufacture, distribution, use and end-of-life options. Specifically, LCA assesses and quantifies all the relevant inputs (e.g. materials, energy, resources) and outputs (e.g. waste, pollution, co-products) for each of these life cycle stages. It thus differs in this way from other similar, single-dimensional tools, such as carbon, water or energy footprinting. The holistic nature of LCA allows identification of burden shifting between impact categories, between life cycle stages or between different locations [8].

LCA is composed of four phases [7]: (1) the goal and scope definition; (2) the life cycle inventory analysis (LCI); (3) the life cycle impact assessment (LCIA); and (4) interpretation.

Goal and scope definition.   This proposes the questions to be answered, the methods to consider, setting the boundaries of the study (i.e. what life cycle stages to include), and choosing the functional unit of the system. The defined goal should answer questions such as 'What is the motivation of the study?' and 'Who is the target audience and how are the final results going to be used?' The scope of the study should include clear system boundaries that make explicit each product or process included in the data collection and environmental assessment. Possibly the most significant output of the goal and scope definition is defining the functional unit. The functional unit identifies the unit to which all environmental impacts are scaled and is generally based on an intended function, particularly in a properly scoped cradle-to-grave assessment. Such a methodology allows any given service (e.g. 'attending a meeting') to be compared and contrasted by different competing life cycle systems (e.g. train, car and videoconferencing). In the case of green solvents, a function could be 'the use of 1 kg of paint', given that solvents are used heavily in the paints industry. The idea of the functional unit can be somewhat non-intuitive at first, but properly identifying the functional unit will ensure the most appropriate interpretation of your results and greater transparency.

Life cycle inventory analysis (LCI).   This involves the generation and/or collection of inputs and outputs (e.g. mass and energy flows) that are used in each process along the life cycle. Data collection is generally the most time-consuming task in an LCA. Many life cycle inventory databases exist, which house an enormous amount of industrial and commercial product and process information. However, in many cases, particularly with new and emerging technologies such as green solvents, end-users of LCA will need to generate novel inventory data. Both qualitative (i.e. for documentation) and quantitative inventory data (i.e. used in the actual assessment) should be collected. Qualitative data are not necessary, particularly in cases of intellectual property, but could include information about the collection process itself (i.e. how the data were generated or compiled), the time-frame of collection, a description of the technology considered, a discussion of data gaps and assumptions, and a description of the overall reliability and quality of the data.

Quantifiable data are what are used directly in the calculations of the impact assessment (see next section) and involve the amount and units of any energy, material, wastes and co-products consumed or produced during a particular process. It is particularly important to pay attention to the production of co-products (i.e. a process that creates not just your product under consideration but others as well), as this can have important consequences on the overall impact results. If, for instance, the production of a solvent that uses 100 MJ of energy during the production process generates a much more profitable co-product used in the cosmetics industry, how should the energy demand of this system be determined? Would it be more correct to partition the energy usage based on the overall mass of product formed, or perhaps based on the relative market price of each product? There is no single correct answer to this, and the most important thing is to include explicit details about how this was considered and what methods were used. ISO 14044 [9] defines a number of procedures that can be used to handle co-products, most notably system expansion and allocation. Allocation is the process of partitioning input and outputs according to a particular defining characteristic shared by each co-product such as mass or market price.

Life cycle impact assessment (LCIA).   This involves the calculation of environmental and human health burdens resulting from the inventory flows identified in the LCI (previous step). Important stages in the LCIA are the selection and identification of the impact categories, classification (i.e. determining which inventory data are applied to which impact categories) and characterization. A number of well-defined and powerful impact assessment methodologies have been developed in North America, Europe and Japan. These generally include a predefined list of impact categories and characterization factors. Impact categories can include midpoint and endpoint indicators. Midpoint indicators refer to a problem-oriented approach and translate impacts into environmental mechanisms such as climate change, acidification or toxicity. Endpoint indicators refer to a damage-oriented approach and translate environmental mechanisms into areas of protection such as human health, ecosystem quality and natural resources (Figure 6.2).

Owing to the increased complexity and number of assumptions that have to be made while extrapolating from midpoint to endpoint indicators, endpoint characterization should be cautioned by also having greater uncertainty. Additionally, non-mandatory transformations of the impact results could include normalization, grouping and weighting, depending on the needs of the study. These are somewhat controversial techniques, generally due to their high subjectivity, and are discussed further in section 6.5.2.

Interpretation.   Finally, interpretation is an ongoing activity that should be performed throughout the course of the LCA to all assessment phases. Interpretation can help identify the most environmentally relevant life cycle processes

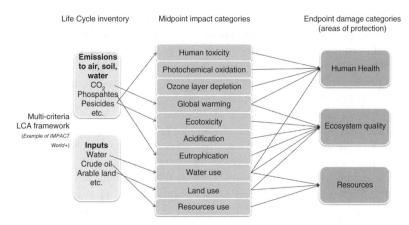

Figure 6.2 Main impact pathways in LCA. Adapted from Impact World+ (http://www .impactworldplus.org/en/index.php).

based on results of the impact assessment, evaluate the completeness, consistency and sensitivity of the information generated/collected, and communicate the applicability and limitations of the study.

6.3 Application of Life Cycle Assessment for Conventional Solvents

LCA is a tool that has been developed particularly with the consideration of industrial chemicals in mind. Well-established life cycle inventories such as ecoinvent [10], the most widely used LCA database, include a large set of inventory data related to the production of conventional and petrochemical solvents [11]. The 50 most important solvents that are currently used in industry are included in this database: alcohols, aldehydes, aliphatic hydrocarbons, aliphatic carboxylic acids, aromatic hydrocarbons, chlorinated hydrocarbons, esters, ethers and glycol ethers, ketones and miscellaneous solvents.

In addition to such databases, specific tools exist such as Ecosolvent, which is designed to compare the life cycle implications of different waste solvent treatment technologies [12]. Using this tool, several organic solvents found in ecoinvent and their related treatment after use (recovery with distillation, incineration, to wastewater treatment plants) can be assessed. Amelio *et al.* [13] propose guidelines for conventional solvent selection, based on results found using the Ecosolvent tool [12]. They classified the solvents into seven different groups depending on their resulting impacts. Solvents such as tetrahydrofuran (THF) were the worst-performing solvents, while methanol was the most benign. This study revealed that the main impacts within the life cycle of a solvent come mainly from its production phase. Incineration as an end-of-life option can be an environmentally preferable option for solvents with a low production-phase environmental impact. On the contrary, solvents with high production-phase impacts should utilize recovery options such as distillation, since it avoids the production of new virgin solvent (and thus the upstream, production-phase impacts).

6.4 Critical Review of Life Cycle Assessment Applied to Green Solvents

In contrast to conventional solvents, green solvents such as bio-based solvents, ionic liquids or supercritical fluids are not explicitly included in databases such as ecoinvent or tools like Ecosolvent. However, the precursors of many of these green solvents, such as CO_2 for use as a supercritical fluid, may be in the database and will be a good starting point for creating the LCI. Yet, LCA has not yet been widely applied to evaluate the environmental profile of potential green solvents.

In this section, we review LCA case studies applied to green solvents that were found from the scientific literature. The identification of studies was broadened to include any explicit use of the term 'green solvent' or a previously identified green solvent category (Table 6.1) or the recovery of conventional solvents.

6.4.1 Criteria of the Review

The selected literature is classified according to criteria related to the four phases of LCA as explained in Table 6.2.

6.4.2 Results of the Review

Overall, we identified seven papers in the scientific literature and one refereed conference poster presentation that provide an LCA involving green solvents or

Table 6.2 *Criteria of the review according to the different LCA phases.*

LCA phase	Criteria
Goal and scope	Goal of the study
	Green solvents assessed/compared
	Conventional solvents assessed/compared
	Level of product development
	Life cycle steps taken into account
	End-of-life scenario
Life cycle inventory	Foreground data source
	Background data source
	LCA software for modelling the inventory
Life cycle impact assessment	LCIA method
	Impacts taken into account
	Normalization, weighting?
	Other criteria non-LCA-related
Interpretation	Sensitivity analysis?
	Main findings

the recovery of conventional solvents. Key results of the review are summarized in Tables 6.3 and 6.4 and are discussed further in the following subsections.

6.4.2.1 Goal

Six out of the eight studies compared a green solvent with conventional solvents [14, 15, 17, 19–21], while the other two studies compared different end-of-life scenarios of conventional solvents for consideration as a 'greener' solvent [16, 18]. Several types of green solvents were studied, including bio-based solvents [21], ionic liquids [14, 15, 17] and supercritical fluids [20].

The applications for which the solvents are used included manufacture of cyclohexane, synthesis of pharmaceutical products, cellulose dissolution, CO_2 capture technologies, production of porous materials and production of nanomaterials.

6.4.2.2 Scope of the Studies

Only half of the studies included a full cradle-to-grave assessment of the solvent, i.e. including the production, use and end-of-life phases. This limitation was cited mainly as a cause of a lack of data for modelling the particular life cycle stages such as the use and end-of-life phases [14–16], although some studies focused specifically on one life cycle stage purposefully due to the intended goals of the study (e.g. to identify best end-of-life practices) [18]. Of the studies that involve the end-of-life phase, two different scenarios were considered: recovery and incineration. Furthermore, the analyses were mostly completed using lab-scale defined data and processes.

6.4.2.3 Life Cycle Inventory

The LCI data used in the studies can be separated between foreground and background data. Foreground data refers to the system of primary concern, i.e. the set of processes whose selection or mode of operation is affected by decisions based on the study [22]. In the case of solvents, it typically includes the quantity of solvents used in the reaction, the recovery rate of the solvent, energy used in the reaction, direct emissions into the air from the use of solvents, etc. Background data refers to all the other processes interacting indirectly with the foreground system, i.e. the production of chemicals, energy, water, infrastructures, etc.

Three main sources of foreground data are used in the studies: experimental data (in four studies), process simulation (in two studies), and generic data from LCA databases (in two studies). Foreground data can also be completed with literature information. As for background data, seven studies use the ecoinvent database whereas one study uses the GaBi database.

All the major LCA software has been used in the reviewed studies: SimaPro is used in four studies, the Ecosolvent tool presented in section 6.3 is used in two studies, openLCA, GaBi and Umberto are each one used in one study.

Table 6.3 *Goal and studied solvents from the literature.*

Citation	Goal	Green solvent studied[a]	Conventional solvents studied[a]
Zhang et al. [14]	Compare solvents for the manufacture of cyclohexane and in a Diels–Alder reaction	Ionic liquid: [Bmim][BF$_4$]	LPDE, water
Reinhardt et al. [15]	Compare solvents for the Diels–Alder reaction of cyclopentadiene and methyl acrylate	Ionic liquid: [C$_6$MIM][BF$_4$], citric acid/N,N-dimethylurea, solvent-free	Methanol, cyclohexane, acetone, methanol/water
Raymond et al. [16]	Study the environmental impact of solvent recovery in the pharmaceutical industry	N.A.	Acetone, acetonitrile, diethyl ether, ethanol, hexane, methanol, THF, toluene
Righi et al. [17]	Compare two solvents for cellulose dissolution processes	Ionic liquid: [Bmim][Cl] and NMMO/H$_2$O	N.A.
Luis et al. [18]	Compare distillation and incineration with energy recovery for the treatment of waste solvent mixtures	N.A.	Mixtures: acetonitrile/toluene, acetonitrile/toluene/THF, ethyl acetate/water and methanol/THF
Fadeyi et al. [19]	Compare two solvents for CO$_2$ post-combustion capture	Advanced solvent (KS-1)	Monoethanolamine
Tsang et al. [20]	Optimization of the production of barium titanate nanomaterials using supercritical fluids and comparison to conventional methods	Supercritical ethanol/water, methanol/water, isopropanol/water	N.A.
Foulet et al. [21]	Compare two solvents for producing emulsion-templated porous material	Vegetable oil: castor oil	1,2-Dichloroethane

[a]Solvent abbreviations: [Bmim][BF$_4$], 1-butyl-3-methylimidazolium tetrafluoroborate; LPDE, lithium perchlorate diethyl ether; [C$_6$MIM][BF$_4$], 1-hexyl-3-methylimidazolium tetrafluoroborate; THF, tetrahydrofuran; [Bmim][Cl], 1-butyl-3-methylimidazolium chloride; NMMO, N-methylmorpholine-N-oxide.

Table 6.4 Key points of the analysis of the reviewed papers.

	Scope				Life cycle inventory				Life cycle impact assessment				
Citation	Life cycle steps of the solvents[a]	EOL scenario for the solvents	Scale of the analysis		Foreground data source	Background data source	Software	LCIA method	Mono/ multicriteria	Impact categories[b]	Normalization and weighting?	Uncertainty analysis?	Other criteria non-LCA
Zhang et al. [14]	P, U	—	Lab scale		Experimental data	ecoinvent 2	SimaPro 7	CML-IA	Multi (11)	A, AD, E, ET(3), GW, HT, OD, POC, VOC	No	No	No
Reinhardt et al. [15]	P, U	Recovery	Lab scale		Experimental data	ecoinvent 2	Umberto	—	Mono (1)	CED	No	No	Environmental health factor
Raymond et al. [16]	P, EOL	Recovery by distillation, incineration,	Industrial scale		ecoinvent 2	ecoinvent 2	Ecosolvent, SimaPro 7	—	Multi (2)	GW, CED	No	No	No
Righi et al. [17]	P, U, EOL	Recovery	Lab scale		Chemical process simulation	GaBi 4 database, literature	GaBi 4	CML-IA	Multi (11)	A, AD, E, ET(3), GW, HT, OD, POC, VOC	No	No	
Luis et al. [18]	EOL	Distillation, recovery, incineration	Industrial scale		Ecosolvent	ecoinvent 2	Ecosolvent	Eco-indicator 99, UBP'97	Multi (2) and single scores	GW, CED	Yes	Yes	No

Fadeyi et al. [19]	P, U, EOL	Recovery	Industrial scale	Process simulation	ecoinvent 2	Simapro 7	CML-IA	Multi (9)	A, AD, E, ET(3), GW, HT, POC	No	No	No
Tsang et al. [20]	P, U, EOL		Lab scale	Experimental data, industrial literature	ecoinvent 2	openLCA 1.4	ReCiPe 2008	Multi (18)	A, AD, E(2), ET(3), GW, HT, OD, POC, LU(3), IR, WD	No	No	No
Foulet et al. [21]	P, U, EOL	Recovery of the conventional solvent	Lab scale	Experimental data, literature	ecoinvent 3	Simapro 8	CML-IA	Multi (11)	A, AD, E, ET(3), GW, HT, OD, POC, LU	No	No	No

[a]Life cycle steps of the solvents: P, production; U, use; EOL, end-of-life.

[b]Impact categories: A, acidification; AD, abiotic depletion; E, eutrophication (can include marine and freshwater); ET, ecotoxicity (can include freshwater, terrestrial and marine); GW, global warming; HT, human toxicity; OD, ozone depletion; POC, photochemical ozone creation; VOC, volatile organic compounds; CED, cumulative energy demand; LU, land use (can include agricultural, natural and urban); IR, ionizing radiation; WD, water depletion.

6.4.2.4   Life Cycle Impact Assessment

Seven out of the eight papers included a multi-criteria analysis whereas Reinhardt *et al.* [15] only considered cumulative energy demand (CED) and thus do not show possible trade-offs to other impact categories. Four studies used CML-IA [22] because it was the most consensual method in the 2000s. However, new methods that are more reliable have been developed since then. However, only a single study [20] used a more recent LCIA methodology, i.e. ReCiPe [23]. One study chose to apply normalization, weighting and grouping of results into a single score [18].

6.4.2.5   Interpretation – Main Findings

Most of the comparative studies show that the presumed green solvents may have higher potential impacts than conventional solvents or may involve burden shifting to other life cycle stages (production or end-of-life) or impact categories. The main conclusions for (1) ionic liquids, (2) supercritical fluids (3) bio-based solvents and (4) end-of-life scenarios of conventional solvents are summarized below.

(1) Zhang *et al.* [14] conclude that replacement of conventional processes by new processes based on green solvents could be environmentally beneficial if the solvent production generates small impacts, if it is reusable and if it results in high yields and easy separation. In their study, the use of ionic liquids does not meet these requirements for the production of cyclohexane because of the high impacts generated during the production. However, the use of the ionic liquids might be environmentally attractive for production of high-value materials (e.g. pharmaceuticals). Reinhardt *et al.* [15] draw the same type of conclusion for ionic liquids and state that these solvents will be attractive if their separation efficiency and their recyclability can be improved. Righi *et al.* [17] found that the use of ionic liquids instead of a conventional process for cellulose dissolution is promising from an environmental point of view but ask for further investigation to confirm the 'greenness' of the new processes.

(2) Supercritical fluids have many potential advantages over conventional use of the same solvents; however, current lab- and pilot-scale technologies make it difficult to fully evaluate whether these advantages will be seen [20]. One of the key aspects of using supercritical fluids is the temperatures and pressures at which they operate. Energy use for increasing solvent temperature to its critical point are generally much more significant than for increasing the pressure. Therefore, depending on the amount of solvent used, impacts could be strongly correlated with a solvent's critical temperature, as some solvents' critical temperatures are an order of magnitude difference – for example, between carbon dioxide (31°C) and water (374°C).

(3) Foulet *et al.* [21] found that the use of castor oil generates much more impacts than petrochemical solvent (dichloroethane) in almost all impact categories even if the use of bio-based solvents decreases the impact on climate change.

One of the key parameters in order to improve the environmental performance of castor oil is improving its separation and its recyclability rate. More generally, bio-based solvents can generate important burden shifting to the impact categories related to agriculture, i.e. (eco)toxicity because of the use of pesticides, eutrophication because of chemical fertilizer and agricultural land occupation.

(4) Regarding the comparison of end-of-life scenarios, Raymond *et al.* [16] have shown that implementing solvent recovery can considerably decrease the impacts compared to incineration and production of new solvents. An important way to improve the environmental performance of solvents is therefore to ensure their recovery. On the other hand, Luis *et al.* [18] have observed that the choice between incineration and recovery of mixtures of solvents depends on the nature of the waste. Energy used for the separation (by distillation) of solvents might be very high and in some cases incineration is environmentally preferable. This is the case, for example, for water–ethyl acetate mixtures.

6.5   Discussion: Methodological Challenges

The use of LCA in the presented case studies has shown its capacity to evaluate trade-offs in environmental impacts for the overall life cycle of solvents.

6.5.1   Life Cycle Inventory Analysis: from Lab, to Pilot, to Industrial Scale

Most LCA studies evaluating new green solvents have been conducted with lab-scale data. There is a need to upscale these results to industrial scale because the experimental conditions are quite different. The extrapolation is a difficult approach that aims a $1/10 \, m^3$ dimensioning of operating systems across several or even a hundred cubic metres. Several studies in the LCA field propose strategies to upscale lab or pilot scale to industry scale [24]. Wider research proposes methodology for the use of LCA as a development tool within early research [25].

LCA results based on lab-scale data can still be used to improve the technology early in the developmental phase, particularly when comparing internal alternative options [20]. However, when compared with alternatives assessed with industrial-scale data, the lab-scale assessment will likely see significant relatively poor performance. Lab-scale operations do not operate with 'economies of scale' conditions, generally resulting in inefficient resource and energy consumption.

6.5.2   Life Cycle Inventory Analysis: Use of Up-to-Date Methods

Mono-criterion approaches such as carbon footprint and energy balances should be avoided in the future since other impact categories such as (eco)toxicity should be evaluated. Also, most recent LCIA methods such as ReCiPe or ILCD [26] should be used instead of outdated methods (CML-IA, Eco-indicator 99, etc.). Regarding the use of a single score in the Ecosolvent tool, this aggregation does not enable one to look specifically at impacts such as toxicity or resource depletion. Also, it is to be noted that the weighting steps and associated single scores are based on value choices and are not scientifically based [7].

For specific impact categories, consensual methods have been developed and should be used in the future, thus increasing the reliability of the impact assessment phase. This is particularly the case for human toxicity and ecotoxicity (e.g. Usetox model from Rosenbaum *et al.* [27]), which are major impacts occurring during the use phase of solvents.

LCIA methods have been developed with broad global, continental and 'regional' spatial and temporal scales. For example, current models may track the acidification potential of a waterway; however, this is generally accomplished using generic landscape descriptions not representative of any specific body of water, while averaging acidification potential over an extended period of time. It would be ideal if all inventory flows could be connected to their impact with the most specified localization (i.e. a specific river) and time (i.e. single-pulse emission of pollutants). However, if such models are to be developed, this would require massive amounts of information to describe all such local landscapes and inventory data. Staying with the example of acidification, data regarding nutrients, biota, geographical information and physical transformations (e.g. type of flow) for all waterways would need to be quantified to obtain localized acidification potential.

New developments of LCIA methods include finer geographical and temporal resolutions. This is the case, for example, for water deprivation [28] or eutrophication [29], which can be useful to increase the relevance of LCA applied for agricultural products (e.g. for bio-based solvents). However, these developments are not yet operational in LCA software, and further developments are needed to improve the resolution of other impact categories such as (eco)toxicity.

Similarly, each inventory item collected would need to include the geospatial and temporal data, which are not representative of the current state of the science. As discussed in the next subsection, coupling LCA with other methods is sometimes needed to tackle these issues.

6.5.3 Coupling Life Cycle Analysis with Other Environmental Assessment Methods

Although LCA is unique in the fact that each assessment tabulates a holistic set of material and energy resources as well as a host of diverse environmental impacts, it may still be necessary to couple LCA with other tools and assessment methodologies, depending on the scope of the study.

It is important to remember that the results of an LCA are relative values. Impacts should not be considered as absolute values that actually estimate the true risk of any given impact. LCA results are intended to be used as either hotspot identification within a single product or in comparison with the results of other options, alternatives, products or processes that fulfil the same functional unit. This is, arguably, a necessary artifact of the LCA methodology itself.

Other tools can be used in combination with LCA in order to enhance the resolution of a particular impact under consideration. Human toxicity and ecotoxicity are two particularly relevant impacts that should be given extra consideration when assessing the nature of a potential green solvent. This is not only because five of the 12 principles of green chemistry deal directly or indirectly with safety and toxicity, but also because of the need to comply with workplace protection protocols or meet regulatory requirements, which is outside the scope of LCA. Potential integration of human health risk assessment (HHRA) and ecological risk assessment (ERA) with LCA is a subject that has been proposed previously for regular chemicals [30, 31] and more recently has received increased attention in view of emerging technologies, such as engineered nanomaterials [32–34]. Emerging technologies catch the attention not only of eager researchers and industries, but also of concerned consumers and cautious government agencies. Therefore, a thorough understanding of potential toxicological impacts from green solvents across the entire life cycle is recommended. LCA can help identify potential hotspots in the life cycle where toxicity may be of greatest concern, such as due to occupational exposures when using the solvent or downstream ecological exposures when solvent waste is disposed of in a landfill. As was discussed previously, such results will be relative, and thus HHRA and ERA can be used in order to provide estimates of absolute risk.

6.5.4 Using Multi-Criteria Decision Approaches for Life Cycle Analysis

In LCA, the end-user is left with many competing impact results to consider. This can complicate the job of the decision-maker if left without a transparent, well-defined methodology to apply in the decision-making process. This can be further complicated if other tools such as HHRA are introduced as discussed above. Using methods such as multi-criteria decision analysis (MCDA) is one approach to assisting the decision-making process [35]. In general, decision analysis tools allow for the explicit weighting, scoring and ranking of competing criteria. Various approaches exist, but optimization (i.e. calculating a single score for each option) models such as multi-attribute value theory and outranking (i.e. calculating the weight of evidence across multiple criterion) models such as Preference Ranking Organization METHod for Enrichment Evaluations (PROMETHEE) are commonly used techniques for MCDA [36].

6.5.5 Broadening the Scope of the Application of Life Cycle Analysis for Solvents

LCA has been used to compare conventional and new solvents in some applications such as the manufacture of chemicals, the capture of CO_2 and the pharmaceutical industry. The use of LCA could be broadened to other sectors that use large quantities of solvents, e.g. in analytical chemistry. A study already provides insight into the environmental evaluation of liquid chromatography and the choice of solvents by using the Ecosolvent tool [37].

6.6 Conclusion

LCA is a well-established methodology to evaluate the environmental performance of solvents in a holistic way. It has already proven its worth for the identification of environmental hotspots in the life cycle of conventional solvents and for the comparison of different scenarios. Its application to new solvents such as ionic liquids, bio-based solvents or supercritical fluids helps to demonstrate the 'greenness' of these solvents. A limited number of LCA papers already studied presumed green solvents, based on different hypotheses and methods as explained in our critical review. Overall, the presumed green solvents may have higher potential impacts than conventional solvents or may involve burden shifting to other life cycle stages (production or end-of-life) or impact categories, especially because of the high impacts generated during their manufacture, or because they are not recoverable. However, LCA helps to identify eco-design solutions in order to improve the environmental performance of these new solvents. Several methodological challenges remain in order to improve the consistency of such environmental assessment, e.g. upscaling inventory data from lab to industrial scale, using up-to-date LCIA methods, or coupling LCA with other methods such as risk assessment.

References

1. Ceresana (2014) Market study: solvents, 3rd edn. http://www.ceresana.com/en/market-studies/chemicals/solvents/ (accessed 1 September 2015).
2. Anastas, P.T. and Warner, J.C. (1998) *Green Chemistry: Theory and Practice*, Oxford University Press, Oxford, p. 30.
3. Capello, C., Fischer, U. and Hungerbühler, K. (2007) What is a green solvent? A comprehensive framework for the environmental assessment of solvents. *Green Chem.*, **9**, 927–934.
4. Clark, J.H. and Tavener, S.J. (2007) Alternative solvents: shades of green. *Org. Process Res. Dev.*, **11**, 149–155.
5. Curzons, A.D., Constable, D.C. and Cunningham, V.L. (1999) Solvent selection guide: a guide to the integration of environmental, health and safety criteria into the selection of solvents. *Clean Technol. Environ. Policy*, **1**, 82–90.
6. Jiménez-Gonzalez, C., Curzons, A.D., Constable, D.J.C. and Cunningham, V.L. (2004) Expanding GSK's Solvent Selection Guide – application of life cycle assessment to enhance solvent selections. *Clean Technol. Environ. Policy*, **7**, 42–50.
7. ISO (2006) *ISO 14040:2006 – Environmental Management – Life Cycle Assessment – Principles and Framework*. International Organization for Standardization, Geneva.
8. Finnveden, G., Hauschild, M.Z., Ekvall, T., et al. (2009) Recent developments in life cycle assessment. *J. Environ. Manag.*, **91**, 1–21.

9. ISO (2006) *ISO 14044:2006 – Environmental Management – Life Cycle Assessment – Requirements and Guidelines.* International Organization for Standardization, Geneva.

10. Frischknecht, R., Jungbluth, N., Althaus, H., et al. (2007) *Overview and Methodology,* ecoinvent Report No. 1. Swiss Centre for Life Cycle Inventories, Dübendorf.

11. Sutter, J. (2007) Life Cycle Inventories of Petrochemical Solvents, ecoinvent Report No. 22. Swiss Centre for Life Cycle Inventories, Dübendorf.

12. Capello, C., Hellweg, S. and Hungerbühler, K. (2006) User Guide. The Ecosolvent Tool. ETH Zurich. https://www1.ethz.ch/sust-chem/tools/ecosolvent/eco_manual/ (accessed 1 September 2015).

13. Amelio, A., Genduso, G., Vreysen, S., et al. (2014) Guidelines based on life cycle assessment for solvent selection during the process design and evaluation of treatment alternatives. *Green Chem.*, **16**, 3045–3063.

14. Zhang, Y., Bakshi, B.R. and Demessie, E.S. (2008) Life cycle assessment of an ionic liquid versus molecular solvents and their applications. *Environ. Sci. Technol.*, **42**, 1724–1730.

15. Reinhardt, D., Ilgen, F., Kralisch, D., et al. (2008) Evaluating the greenness of alternative reaction media. *Green Chem.*, **10**, 1170–1181.

16. Raymond, M.J., Slater, C.S. and Savelski, M.J. (2010) LCA approach to the analysis of solvent waste issues in the pharmaceutical industry. *Green Chem.*, **12**, 1826–1834.

17. Righi, S., Morfino, A., Galletti, P., et al. (2011) Comparative cradle-to-gate life cycle assessments of cellulose dissolution with 1-butyl-3-methylimidazolium chloride and *N*-methyl-morpholine-*N*-oxide. *Green Chem.*, **13**, 367–375.

18. Luis, P., Amelio, A., Vreysen, S., et al. (2013) Life cycle assessment of alternatives for waste-solvent valorization: batch and continuous distillation vs incineration. *Int. J. Life Cycle Assess.*, **18**, 1048–1061.

19. Fadeyi, S., Arafat, H.A. and Abu-Zahra, M.R.M. (2013) Life cycle assessment of natural gas combined cycle integrated with CO_2 post combustion capture using chemical solvent. *Int. J. Greenh. Gas Control*, **19**, 441–452.

20. Tsang, M., Sonnemann, G.W., Philippot, G. and Aymonier, C. (2014) Supercritical and sustainable synthesis of nanoparticles. Poster presented at *EcoBalance Conf.*, Tsukuba, October.

21. Foulet, A., Birot, M., Sonnemann, G. and Deleuze, H. (2015) Life cycle assessment of producing emulsion-templated porous materials from Kraft black liquor – comparison of a vegetable oil and a petrochemical solvent. *J. Clean. Prod.*, **91**, 180–186.

22. Guinée, J.B., Gorrée, M., Heijungs, R., et al. (2002) *Handbook on Life Cycle Assessment: Operational Guide to the ISO Standards.* Kluwer Academic, Dordrecht.

23. Goedkoop, M., Heijungs, R., Huijbregts, M., et al. (2009) *ReCiPe 2008: A Life Cycle Impact Assessment Method* . . . . Ministry of Housing, Spatial Planning and Environment, The Netherlands.

24. Shibasaki, M., Fischer, M. and Barthel, L. (2007) Effects on life cycle assessment – scale up of processes. In *Advances in Life Cycle Engineering for Sustainable Manufacturing Businesses.* Springer, London, pp. 377–381.

25. Hetherington, A.C., Borrion, A.L., Griffiths, O.G. and McManus, M.C. (2014) Use of LCA as a development tool within early research: challenges and issues across different sectors. *Int. J. Life Cycle Assess.*, **19**, 130–143.

26. EC JRC-IES (2010) *ILCD Handbook: Analysis of Existing Environmental Impact Assessment Methodologies for Use in Life Cycle Assessment.* European Commission, Joint Research Centre, Institute for Environment and Sustainability.

27. Rosenbaum, R.K., Bachmann, T.M., Gold, L.S., et al. (2008) USEtox – the UNEP-SETAC toxicity model: recommended characterisation factors for human toxicity and freshwater ecotoxicity in life cycle impact assessment. *Int. J. Life Cycle Assess.*, **13**, 532–546.

28. Kounina, A., Margni, M., Bayart, J.-B., et al. (2012) Review of methods addressing freshwater use in life cycle inventory and impact assessment. *Int. J. Life Cycle Assess.*, **18**, 707–721.

29. Cosme, N.M.D., Larsen, H.F. and Hauschild, M.Z. (2013) Endpoint characterisation modelling for marine eutrophication in LCIA. Presented at *SETAC Europe 23rd Annual Meeting*, Glasgow.

30. Bare, J.C. (2006) Risk assessment and life-cycle impact assessment (LCIA) for human health cancerous and noncancerous emissions: integrated and complementary with consistency within the USEPA. *Hum. Ecol. Risk Assess., Int. J.*, **12**, 493–509.

31. Olsen, S.I., Christensen, F.M., Hauschild, M., et al. (2001) Life cycle impact assessment and risk assessment of chemicals – a methodological comparison. *Environ. Impact Assess. Rev.*, **21**, 385–404.

32. Grieger, K., Laurent, A., Miseljic, M., et al. (2012) Analysis of current research addressing complementary use of life-cycle assessment and risk assessment for engineered nanomaterials: have lessons been learned from previous experience with chemicals? *J. Nanoparticle Res.*, **14**, 1–23.

33. Shatkin, J.A. (2008) Informing environmental decision making by combining life cycle assessment and risk analysis. *J. Ind. Ecol.*, **12**, 278–281.

34. Steinfeldt, M., von Gleich, A., Petschow, U. and Haum, R. (2007) *Nanotechnologies, Hazards and Resource Efficiency*. Springer, Berlin.

35. Linkov, I. and Seager, T.P. (2011) Coupling multi-criteria decision analysis, life-cycle assessment, and risk assessment for emerging threats. *Environ. Sci. Technol.*, **45**, 5068–74.

36. Kiker, G.A., Bridges, T.S., Varghese, A., et al. (2005) Application of multicriteria decision analysis in environmental decision making. *Integr. Environ. Assess. Manag.*, **1**, 95–108.

37. Gaber, Y., Törnvall, U., Kumar, M.A., et al. (2011) HPLC-EAT (environmental assessment tool): a tool for profiling safety, health and environmental impacts of liquid chromatography methods. *Green Chem.*, **13**, 2021–2025.

7

Alkylphenols as Bio-Based Solvents: Properties, Manufacture and Applications

Yuhe Liao, Annelies Dewaele, Danny Verboekend, and Bert F. Sels

Centre for Surface Chemistry and Catalysis, KU Leuven, Heverlee, Belgium

7.1 Introduction

Owing to global concerns about climate change and our heavy dependence on oil, today's society is moving from a fossil-based economy to a sustainable economy. The energy from fossil oil should eventually be replaced by alternative energy sources, like solar, wind and geothermal. However, accordingly, also the supply of oil-derived chemicals will vanish, meaning that alternative sources for base chemicals should be identified. Biomass, as a renewable carbon source, has received increasing interest to produce a variety of platform chemicals and many downstream end products, and is globally available [1, 2].

Major potential biomass streams are lignocellulose, glycerides from animal fats and microalgae. Among these, glycerides and microalgae are used to produce bio-diesel [3]. In the conversion of glycerides to diesel, glycerol is produced as a side product. Lignocellulosic biomass, containing sugar-based materials (hemicellulose and cellulose) and lignin, is the most abundant terrestrial biomass resource. A number of technologies have been developed to convert lignocellulose into

Bio-Based Solvents, First Edition. Edited by François Jérôme and Rafael Luque.
© 2017 John Wiley & Sons Ltd. Published 2017 by John Wiley & Sons Ltd.

important platform compounds. For example, the conversion of hemicellulose and cellulose to furfural (FuAl) and 5-hydroxymethylfurfural (HMF), respectively, has been widely studied [4, 5]. These platform chemicals can be further upgraded into fuel additives and base chemicals.

Some biomass-derived chemicals have been proposed as bio-based green solvents for catalysis, organic reactions, separations and materials chemistry because of their renewability, which meets the requirements of green chemistry (Figure 7.1). Typically, glycerol from glycerides, being a polyol, has been widely reviewed and used as a green solvent in many organic reactions or other fields [6–9]. With respect to lignocellulose, some derived chemicals such as 2-methyltetrahydrofuran (2-MeTHF) [10], gamma-valerolactone (GVL) [7, 11, 12], lactic acid [13] and ethyl lactate (EL) [14] have also been reported as novel solvents to enhance the reaction rate and product selectivity of many organic reactions. Especially, GVL has been widely proposed as a novel and green solvent for the conversion of hemicellulose and cellulose. The synthesis routes and applications of these solvents have been discussed as well [8–12, 14, 15].

In addition to these non-aromatic chemicals, aromatic chemicals can be produced from lignin as a result of its aromatic-containing structure (Figure 7.1) [16, 17]. Particularly, the short-chain alkylphenols can be manufactured directly from lignin by using recently developed technologies [16, 17]. The long-chain alkylphenols can also be produced from biomass in a few steps. However, thus far,

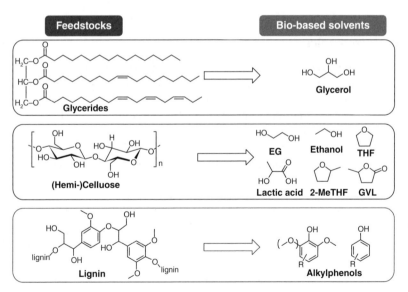

Figure 7.1 Bio-based solvents manufactured from biomass by using bio- and chemocatalysis. EG, ethylene glycol; THF, tetrahydrofuran; 2-MeTHF, 2-methyltetrahydrofuran; GVL, gamma-valerolactone. (A colour version of this figure appears in the plate section.)

limited studies have focused on using alkylphenols as bio-based solvents. Hence, this chapter will discuss the main properties of alkylphenols, their production and their utilization as solvents. Finally, the stability and toxicity aspects of alkylphenols are highlighted and an outlook is provided.

7.2 Properties of Alkylphenols

Alkylphenols are a family of phenol-derived organic compounds obtained by replacing one or more of the ring hydrogens with alkyl groups [18, 19]. In this chapter, we mainly discuss the mono-alkylated phenols. Pure alkylphenols appear colourless or white to a pale yellow, and have a phenol-type odour. They show the same sensitivity to oxygen as phenol, in the sense that oxidation can cause discoloration. The properties of alkylphenols are strongly affected by the size and the configuration of the alkyl group. The physical form of alkylphenols at 25°C is determined by the type of alkyl group and the mutual positions of the alkyl and hydroxyl groups. For example, *ortho*-cresol and phenol are solids, whereas *meta*-cresol is a liquid. The boiling points of alkylphenols increase continuously with increasing size of the alkyl group, hence molecular weight (MW) (Figure 7.2). The *para*-alkylphenols have higher boiling points than the corresponding *ortho*-isomers. However, the difference in boiling points between the *meta*- and *para*-isomers is very small, which leads to difficulties in separating them by distillation. The *n*-alkylphenols have higher boiling points than the

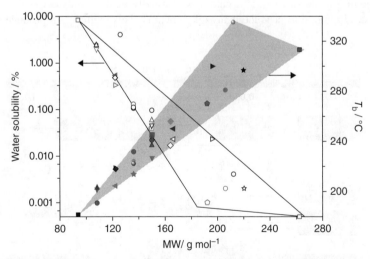

Figure 7.2   The water solubility and boiling point (T_b) of selected mono-alkylated phenols as a function of molecular weight (MW) [20–32]. The hollow symbols represent the water solubility and the filled symbols represent the boiling points of selected mono-alkylated phenols at 25°C. (A colour version of this figure appears in the plate section.)

corresponding *iso*-alkylphenols. The solubility of alkylphenols in water decreases logarithmically with increasing size of the alkyl group (Figure 7.2), which allows the use of alkylphenols as organic solvents. Compared to alkylphenols, the corresponding methoxylated alkylphenols have higher boiling points and lower solubilities in water. The chemical properties of alkylphenols are given in two encyclopaedias [18, 19].

7.3   Manufacture of Alkylphenols

Feedstocks for producing alkylphenols include raw oil, coal and lignocellulose. Several different technologies are utilized for these substrates, which differ substantially as a result of their distinct oxygen contents (Figure 7.3). It is impossible to produce alkylphenols from oil directly due to the absence of oxygen in oil. Accordingly, oxidation is a necessary step for producing alkylphenols from oil. During the gasification of coal, syngas and coal tar are obtained simultaneously. Phenolics are a significant part of coal tar. Phenol and some alkylphenols can be separated from this mixture [19]. Yielding alkylphenols from raw lignocellulose is the third pathway. Owing to the high oxygen content of lignocellulose, it should be subjected to a reduction step. In particular, lignin can be selectively converted into alkylphenols, and the main alkylphenols are short-chain alkylphenols (alkyl chain between C_1 and C_3). Very recently, it has been demonstrated that phenol can be selectively yielded from raw and fossilized (coal) lignocellulose by means of dealkylation [33]. Alkylation of phenol with alkylating agents (alcohols or olefins) in the presence of catalysts (normally acidic catalysts) is the main industrial method to produce many alkylphenols. This method can produce both

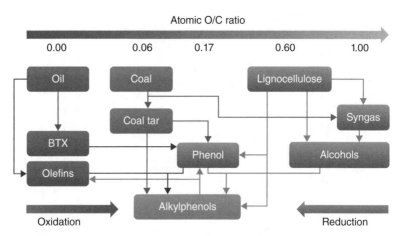

Figure 7.3   Methods to produce alkylphenols from different substrates. BTX, benzene, toluene and xylene. (A colour version of this figure appears in the plate section.)

short-chain (C_1–C_3) and long-chain (C_4–C_{12} alkyl chain length) alkylphenols, but the chain length depends on the alkylating agent.

7.3.1 Oil-Derived Synthesis

The synthesis of alkylphenols from oil includes three steps: (1) producing benzene and alkylating agents (olefins) from raw oil; (2) producing phenol from benzene; and (3) alkylation of phenol with alkylating agents. The cumene hydroperoxide route is the industrial technology for phenol production from benzene [34]. As alkylation is the only way to produce alkylphenols from oil, alkylation of phenol with alkylating agents is discussed here. Many acid catalysts have been used as Friedel–Crafts catalysts for the alkylation of phenol with alkylating agents (normally alkenes or alcohols). The general reaction pathway of Friedel–Crafts alkylation is that an alkyl cation, formed by the interaction of catalysts and alkylating agents, interacts with the nucleophile to produce mono-C-alkylated products or poly-C-alkylated products. Besides these, O-alkylation of the hydroxyl group of phenol also occurs, which produces ether. Except for the O-alkylation reaction, other side reactions such as transalkylation, dealkylation and the side reactions of the alkylating agents also have a great effect on the selectivity of the desired alkylphenols.

For the sake of selectively producing individual alkylphenols through alkylation, the influence of several factors needs to be understood. As shown in Figure 7.4, the process is controlled by three key parameters: (1) reaction conditions, (2) alkylating agents and (3) catalysts. The reactor type (batch vs.

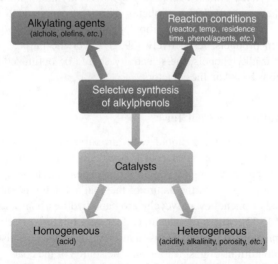

Figure 7.4 Factors that control the selectivity of desired alkylphenols in alkylation of phenol with alkylating agents. Temp., temperature.

continuous reactor), temperature and residence time are important criteria, which impact the alkylation process greatly. A high reaction temperature can lead to side reactions such as dealkylation and transalkylation. A long residence time can cause the formation of multi-alkylated phenols in the product stream [35, 36]. The molar ratio of phenol to alkylating agents is also a parameter that influences the product distribution. Multi-substitution products may be formed when high molar ratios of alkenes to phenol are used [37]. Next, the choice of the alkylating agents is vital due to their different reactivities [38]. In the case of alkylation processes catalysed by heterogeneous bases, the isomerism of the alkylating agent directs the resulting type of alkylphenol [39]. With tertiary olefins as the alkylating agent, the formation of *para*-substituted structures is favoured, while secondary or primary olefins rather favour the formation of *ortho*-substituted alkylphenols. Finally, the catalysts used in this process are of importance [38]. The product selectivity of a homogeneous catalysed process is normally thermodynamically controlled, which presents a separation problem, because the products are a mixture of *para*-, *meta*- and *ortho*-isomers. Compared to homogeneous acid-catalysed processes, the properties of heterogeneous catalysts have more complex effects on the product distributions, and also impact the stability of solid catalysts. For example, zeolites (crystalline acidic aluminosilicates) have been widely used in the alkylation of phenol with olefins or alcohols [37, 40–45]. The ordered micropores enable one to synthesize alkylphenols shape-selectively by tuning the topology of the zeolites and the acidity [46–51]. In addition, the acidity or alkalinity of heterogeneous catalysts influences the selectivity of products and the conversion of phenol. For example, methylation of phenol over heterogeneous basic catalysts yields almost exclusively *ortho*-C-alkylation product (*o*-cresol) [52–56]. Furthermore, alkylation of phenol with 1-propanol and 2-propanol over heterogeneous bases produces 2-*n*-propylphenol and 2-*iso*-propylphenol as the main alkylation products, respectively [39, 57]. For the purpose of producing only the desired alkylphenols, the selectivity should be optimized by inhibiting the side reactions based on these factors.

7.3.2   Separation from Coal Tar

Coal tar is a brown or black liquid with an extremely high viscosity, and is obtained from the gasification of coal [58–60]. In general, the properties and composition of coal tar are highly complex, and depend on the temperature of the thermal degradation process, the nature of the coal and the type of reactor. Coal tar is composed of phenolics, polycyclic aromatic hydrocarbons and heterocyclic compounds [61, 62], where the phenolics feature a mixture of primarily mono- and dihydric phenols [63, 64]. The distribution of these phenols also depends on the operation conditions of gasification and the source of the coal.

Phenols can be easily extracted from coal tar solutions by using aqueous sodium hydroxide due to their weak acidic nature compared to other less polar compounds.

This method is used to recover phenol and cresols from coal tar [65–67]. The extracted phenol mixture undergoes distillation to produce specific alkylphenols. However, because the boiling points of the alkylphenols are very similar, it is not straightforward to separate alkylphenols in pure form by distillation [19]. The distillation product obtained at a certain temperature is therefore a mixture of several alkylphenols. Consequently, other methods should be used to further purify the alkylphenols. For example, separation methods like esterification with oxalic acid and selective dealkylation over specific catalysts and systems have been adopted for separating the *meta-* and *para-*isomers [19, 68].

7.3.3 (Methoxylated) Alkylphenols from Lignin

Lignin is the only renewable polymer that is composed of aromatics. The most important step for producing chemicals from lignin is depolymerization. There are several routes able to convert lignin into phenolic compounds, with the most important being base-catalysed depolymerization, hydrogenation, transfer hydrogenation, pyrolysis and oxidation (Figure 7.5) [16, 17, 69–71]. Through oxidation

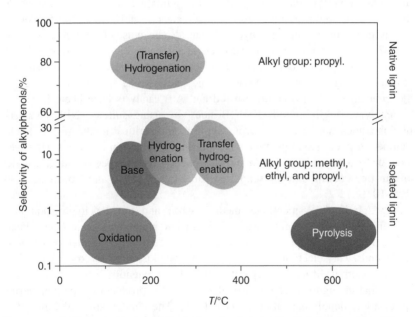

Figure 7.5 Catalytic conversion of lignin into (methoxylated) alkylphenols using different approaches. The typical catalysts used for these processes are as follows. Oxidation: CuO, TiO_2, $LaMnO_3$, $H_3PMo_{12}O_{40}$, etc. Pyrolysis: none, zeolite, etc. Base (base-catalysed depolymerization): NaOH, KOH, NiO-hydrotalcite, etc. Hydrogenation: Ni/C, NiMo, Ru/C, NiRu, etc. Transfer hydrogenation: Pt/Al_2O_3, MoC_{1-x}/AC, Cu-PMO (porous metal oxides), etc. Isolated lignin includes Kraft lignin, Organosolv lignin, soda lignin, steam-exploded lignin, acid-hydrolysed lignin, etc. Native lignin is the lignin of raw lignocellulose, which is also called protolignin.

of lignin, the main products are phenolic aldehydes and acid [72–74]. Pyrolysis of lignin produces a high yield of solid residue (char), and a low liquid product yield is obtained [69, 75]. The yield and selectivity of (methoxylated) alkylphenols obtained from lignin through oxidation and pyrolysis are very low (Figure 7.5).

Base catalytic processes are usually performed in water or alcohols with homogeneous bases, such as NaOH or KOH [76–86]. Aside from these base catalysts, some chemicals such as boric acid and phenol can be added to the reaction system as capping agents to enhance the yield of monomers [82, 84]. Although homogeneous bases show promise for lignin degradation, their recovery and recycling are difficult, especially in water. More recently, it has been reported that it is possible to depolymerize lignin with heterogeneous bases, like nickel on hydrotalcite [87, 88]. The transfer hydrogenation process involves the reductive depolymerization of lignin in the absence of externally provided hydrogen. This process is also referred to as liquid-phase reforming. The hydrogen needed for the reaction is provided by a solvent like 2-propanol or methanol, or by an additive like formic acid. When using solvents as the hydrogenous source, catalysts like copper-doped porous metal oxides catalyse hydrogen transfer from methanol and ethanol to isolated lignin [89–94]. Except for pure alcohol, a mixture of water and ethanol has also been used to convert lignin over Pt/Al_2O_3 and co-catalyst H_2SO_4 [95]. Formic acid as an alternative source of hydrogen has also been used for lignin conversion [96, 97]. In contrast to transfer hydrogenation, standard hydrogenation of lignin involves the conversion of lignin in the presence of external hydrogen over a hydrogenation catalyst. A series of monometallic catalysts like Ni on activated carbon and bimetallic NiM (M = Ru, Rh, Pd and Au) catalysts have been developed to depolymerize lignin [98–103]. Besides, by using sulfided NiW catalysts, alkylphenolics can be isolated from lignin in supercritical methanol [104]. The majority of hydrogenation studies have been performed in a solvent. However, catalytic hydrogenation of lignin into alkylphenols has also been conducted under solvent-free conditions over supported Ru, Pd, Cu, NiMo and CoMo catalysts [105–107].

Although the above-mentioned methods – base and (transfer) hydrogenation – can efficiently depolymerize isolated lignin into monomers or low-molecular-weight fractions, the product mixture consists of a large number of phenolic compounds. The selectivity of a specific compound is relatively low (Figure 7.5), which renders it difficult to separate targeted alkylphenols. It has been demonstrated that the lignin structure strongly impacts the product distributions during the catalytic depolymerization process [101]. The condensation of lignin during isolation or pre-treatment heavily influences the obtained alkyl group of alkylphenols, which results in a mixture of methyl, ethyl and propyl groups. For the purpose of alkylphenol production, the structure of lignin should be preserved as much as possible before depolymerization to attain a high yield and selectivity of monomers. Several studies have reported the conversion of the native lignin of raw biomass over hydrogenation catalysts in aqueous phase. The main products of native lignin degradation are a few monomers (mainly **1, 2, 3** and **4** in Figure 7.6) [108, 109]. Although these studies have shown that high yields and selectivities of these monomers can be attained, the transformation of cellulose and hemicellulose is not discussed. In a report on the one-pot conversion of raw woody biomass into chemicals over a supported carbide catalyst [110], phenolic monomers (**1, 2, 3** and **4** in Figure 7.6) are obtained from lignin and (hemi)cellulose is converted into diols in water.

Some recent studies have reported the reductive fractionation of raw lignocellulose, in which lignin is converted into lignin oil (rich in monomers). Cellulose and hemicellulose are left as a carbohydrate-rich pulp, which can be further upgraded into platform chemicals over bio- or chemocatalysts. Typically, raw biomass is processed in methanol with hydrogenation catalysts like Ru/C and Ni/C in the presence or absence of external hydrogen [111–114]. The main monomers are **1, 2, 3** and **4** (Figure 7.6) with a very high selectivity (Figure 7.5). It has been shown that in this process lignin is first degraded into smaller lignin species consisting of several benzene rings, followed by hydrogenolysis into monomers [111]. Later, it was found that this process is strongly affected by the type of catalyst, the reaction solvent and the reaction conditions [115–117]. Likewise, transfer hydrogenation of raw biomass in a mixture of water and 2-propanol over Raney nickel catalyst gives similar results [118]. Interestingly, a tandem Organosolv pulping (to solubilize lignin over organic solvents) and palladium-catalysed transfer hydrogenation route has been designed to depolymerize native lignin in a mixture of water and ethanol. This process generates 4-propenylguaiacol (*iso*-eugenol) and 4-propenylsyringol (**5** and **6** in Figure 7.6) as the main products. This study suggests that the species liberated from lignocellulose during this process, like formic

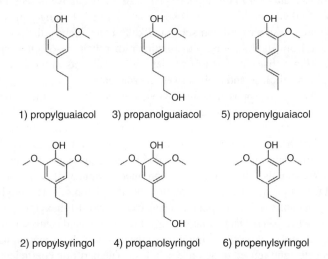

1) propylguaiacol   3) propanolguaiacol   5) propenylguaiacol

2) propylsyringol   4) propanolsyringol   6) propenylsyringol

Figure 7.6   The main monomers from hydrogenation of native lignin.

acid, can act as a hydrogen donor [119]. Apart from propylphenols, ethylphenols were also selectively produced from native lignin [120, 121].

As lignin-derived monomers normally contain one or two methoxyl groups, the methoxyl groups have to be removed to upgrade them into alkylphenols, while preserving the alkyl and hydroxyl groups on the monomers. Although a lot of studies use guaiacol as a model substrate [122], a few studies focused on the conversion of alkylguaiacols like 4-propylguaiacol into alkylphenols. The catalytic hydrogenation of 4-propylguaiacol was studied in a batch and fixed-bed reactor over γ-Al$_2$O$_3$-supported sulfided CoMo, Mo, MoS$_2$, MoS$_2$/NiS and NiMo catalysts in hydrogen atmosphere, with n-propylphenol as the main product [123–125]. The acidic sites of the support were found to promote dealkylation of 4-propylguaiacol into a number of alkylated phenols like ethylphenol. Alkylsyringols like propylsyringol are the most abundant lignin monomer from most lignocellulose, but their conversion into alkylphenols has not been reported yet. The same applies to lignin-derived monomers with propanol and propenyl groups.

7.4   Alkylphenols as Solvent

The production of chemicals and fuels from cellulose and hemicellulose relies heavily on acid-catalysed hydrolysis and dehydration to produce reactive intermediates by removing oxygen. These processes normally proceed in aqueous phase, but the selectivities of the products are often negatively influenced by the high reactivity of the substrate or the targeted products. In order to address this problem, an aqueous–organic biphasic system can be employed. In this case, the targeted products or highly reactive substrates extract into the organic phase to avoid the side reactions that occur in the aqueous phase. As shown above in section 7.2, the solubility of alkylphenols in water is very low, and their boiling points are very high compared to other common solvents like toluene. Hence, the products can be separated from alkylphenols by distillation rather than by solvent evaporation. Figure 7.7 shows that alkylphenols can be used as solvents in the conversion of cellulose, hemicellulose and their derived intermediates.

Alonso *et al.* have proposed using alkylphenols as solvents in the conversion of cellulose and corn stover into gamma-valerolactone (GVL) [126, 127]. This process involves three steps: (1) deconstruction of cellulose into levulinic acid and formic acid in an aqueous H$_2$SO$_4$ solution; (2) liquid–liquid extraction of levulinic acid and formic acid by alkylphenols; and (3) hydrogenation of levulinic acid to GVL in alkylphenols. The levulinic acid produced from cellulose is successfully extracted from the aqueous H$_2$SO$_4$ solution by alkylphenols. Three alkylphenols, 2-*sec*-butylphenol (SBP), 4-*n*-pentylphenol (NPP) and 4-*n*-hexylphenol (NHP), are used as extraction solvents. Interestingly, the sulfuric acid catalyst is not found in the alkylphenol stream, which implies that the aqueous phase containing H$_2$SO$_4$ can be recycled and reused to degrade cellulose. The partition coefficient for levulinic acid extraction with SBP is approximately 2. With increasing alkyl chain

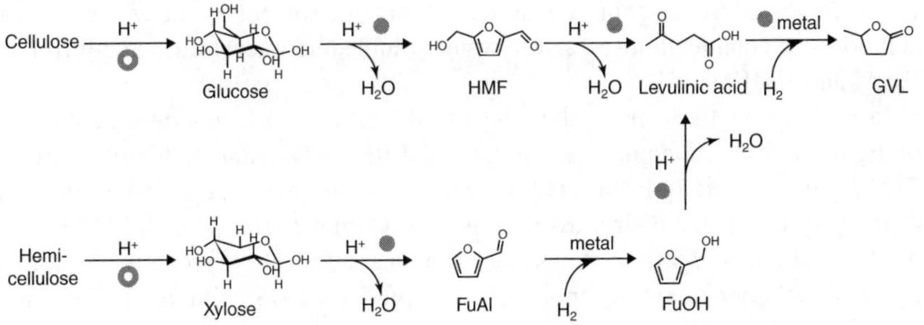

Figure 7.7 Alkylphenols used as solvents in the conversion of biomass. The circles represent the alkylphenols used in these steps. Solid circle: alkylphenols as solvents for this reaction. Hollow circle: alkylphenols as solvents to recycle the solvent used in this reaction. FuAl, furfural; FuOH, furfuryl alcohol; HMF, 5-hydroxymethylfurfural; GVL, gamma-valerolactone.

length of the alkylphenol, the polarity of the alkylphenol is reduced and the partition coefficient for levulinic acid decreases to 1.2 by using NPP, and to 0.8 by using NHP. After extraction of levulinic acid, hydrogenation of levulinic acid to GVL is performed in alkylphenol. GVL is more hydrophobic than levulinic acid and accordingly yields a higher partition coefficient, which enables the alkylphenols to re-extract levulinic acid after hydrogenation. Finally, a higher GVL concentration is obtained. By modifying the catalyst, hydrogenation of the C=C bonds of alkylphenols can be avoided during hydrogenation of the C=O bond of levulinic acid [128]. Following this work, it was demonstrated that hemicellulose can be converted into FuAl and levulinic acid in a biphasic reactor, using alkylphenols as organic solvents [129–131]. In this process, the FuAl and furfuryl alcohol (FuOH) are successfully extracted from the acidic aqueous solutions. Three organic solvents, SBP, NHP and 4-propylguaiacol (PG, a monomer of lignin), are demonstrated to be effective extraction agents for FuAl, FuOH and levulinic acid. In particular, SBP shows an exceptionally high partition coefficient of *ca.* 50 for FuAl in the absence of salt in the aqueous phase. The partition coefficient increases to >90 when the aqueous phase is saturated with NaCl. Besides, only negligible amounts of Cl$^-$ are found in SBP. The FuAl can be distilled from the solvent because of the high boiling point of SBP. The partition coefficient of SBP for FuAl has been demonstrated to decrease with increasing temperature of a SBP/water biphasic system [132]. Hydrolysis of FuOH to levulinic acid has also been conducted in an SBP/water system. The FuOH remains mostly in the SBP phase due to the high partition coefficient (7.5) of FuOH in this biphasic system, which inhibits the side reactions of FuOH (particularly oligomerization) in aqueous phase. Accordingly, the selectivity of levulinic acid is improved by decreasing the concentration of FuOH in acidic aqueous phase. HMF with a yield of 62% from glucose has been achieved in an SBP/water biphasic reaction, and 97% of the produced HMF was extracted with the organic solvent [133, 134]. PG, which can be produced from lignin, was used as an organic solvent for the conversion of cellulose- and hemicellulose-derived water-soluble oligosaccharides into furfurals (FuAl and HMF) [135].

In addition to these pure alkylphenols, Dumesic *et al.* reported on the use of lignin-derived monomers as solvents for the conversion of biomass [109]. The lignin-derived solvents (LDS) mainly comprise 4-propylguaiacol and 4-propylsyringol, and their mass ratio is 1 : 4. Although the yield of LDS in this work is quite low (7 wt.%), it has been demonstrated that the recently developed lignin depolymerization technologies can produce LDS with high yield and selectivity. The LDS show similar partition coefficients to PG for extracting FuAl, FuOH, HMF and levulinic acid in aqueous acid solutions, due to the high structural similarity between 4-propylguaiacol and 4-propylsyringol (Figure 7.6; the difference between them is only one methoxyl group) even though their mass ratio is 1 : 4. In all examined reactions – glucose to HMF, xylose to FuAl, HMF to levulinic acid, and FuOH to levulinic acid – LDS exhibits similar performance to PG, and can be compared with the results of SBP.

Alkylphenols are not only useful as reaction solvents, but they can also be employed to recover the reaction solvent in biomass conversions. By applying a homogeneous mixture of GVL and water as a solvent, a non-enzymatic biomass depolymerization process was developed to produce sugars [136]. From an economic perspective, the key step in this process is to separate the GVL from the aqueous phase and reuse it. It has been demonstrated that GVL can be separated from water by addition of liquid CO_2. However, the high pressure of CO_2 in this separation process might cause safety issues and increase both equipment and energy costs. With the aim of solving this, alkylphenols like SBP, nonylphenol (NNP), *tert*-butylphenol (TBP) and PG have been utilized as solvents to extract GVL from the acidic aqueous phase. The results reveal that alkylphenols can be effective phase modifiers to render the GVL insoluble in the aqueous phase [137, 138]. Especially, in the case of NNP, the extracted aqueous sugar solution is compatible with biological upgrading. Furthermore, no dilution is required for robust growth, and high ethanol yield is obtained.

Because a lot of organic solvents can serve as extraction solvents, efficient methods should be established to determine the optimal solvents. Palkovits *et al.* utilized a computational method to screen and identify solvents for HMF extraction from an aqueous phase [139]. From the extensive list of solvents studied (Figure 7.8), it was found that a number of short-chain alkylated phenols like *ortho*-ethylphenol and *ortho-tert*-butylphenol can effectively extract HMF from aqueous fructose solution due to the high HMF partition coefficients. SBP is also identified by this method, but only in conjunction with the addition of NaCl. Because it is inevitable to extract fructose while extracting HMF, two alkylphenols, which have high partition coefficients for HMF and low extraction

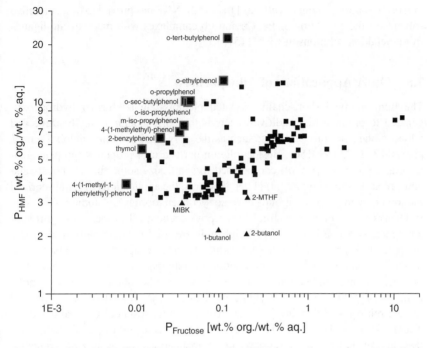

Figure 7.8 The 110 feasible solvents for extracting HMF from aqueous phase [139]. Each point represents a solvent: the triangles are the solvents used for verification of the computational method (COSMO-RS); the squares are the solvents identified through COSMO-RS to show improved extraction performance. The larger marked square points are alkylphenols. Adapted from Blumenthal *et al.* (2016) *ACS Sustain. Chem. Eng.*, **4**, 228–235. Reproduced with permission of the American Chemical Society. (A colour version of this figure appears in the plate section.)

for fructose, were selected for experimental verification. Thus *ortho*-propylphenol and *ortho-iso*-propylphenol were selected and identified as efficient solvents for HMF extraction. The economics of alkylphenols as solvents in biomass conversion have also been investigated [140–143]. The techno-economic analysis shows that using alkylphenols as solvents to produce fuels from (hemi)cellulose is an economically attractive alternative.

Alkylphenols are also used as solvents in other aspects. *p*-Cresol was reported as a solvent in lignin depolymerization by which complete suppression of char formation is realized. The lignin-derived intermediates interact with *p*-cresol in that process [144, 145]. Phenol derivatives, such as SBP and NNP, were also explored for the liquid–liquid extraction of 2,3-butanediol from aqueous solutions [146]. In other contexts, *m*-cresol is used as a solvent for preparing nanostructured SnO_2

thin films and for polymer synthesis [147–150]. NNP has proved to be an effective solvent for the extraction of Fe, Cu and Pb complexes with macrocyclic ligands from metal ion solutions in water [151, 152].

7.5   Other Applications of Alkylphenols

The lignin-derived short-chain alkylated phenols can be further hydrodeoxygenated to produce fuel additives, such as the corresponding alkylbenzene and alkylcyclohexane, or chemicals, such as the polymer precursor cyclohexanones [108, 112, 118, 153–156]. Besides, lignin-derived aryl propenes (propenylguaiacol and propenylsyringol) can be used to synthesize some fine chemicals and new green plastics (Figure 7.9) [119]. However, the applications of lignin-derived monomers with propanol groups have not yet been thoroughly studied.

Thanks to the great versatility in the physicochemical properties of different alkylphenols, some long-chain alkylphenols are used commercially in a variety of applications. Nonylphenol (NNP), which used to occupy the largest part of the market, is mainly used in the production of nonylphenol ethoxylates as surfactant, with minor applications in the production of phenolic resins and phenolic oximes [157, 158]. Smaller alkylphenols (C_5–C_8 alkyl groups) are also converted to their ethoxylated derivatives for emulsifying purposes, but to a smaller extent. These alkylphenols are mainly used to produce plastic additives, lubricant additives and phenolic resins, and have minor applications as vulcanizing agents for rubber and chain terminators for polycarbonates [159, 160]. The major use of 4-dodecylphenol is quite different from those of other alkylphenols since it has a poor surfactant performance. The great majority of this alkylphenol is used in the production of oil and lubricant additives by converting it to calcium alkyl phenolate sulfides [31, 161].

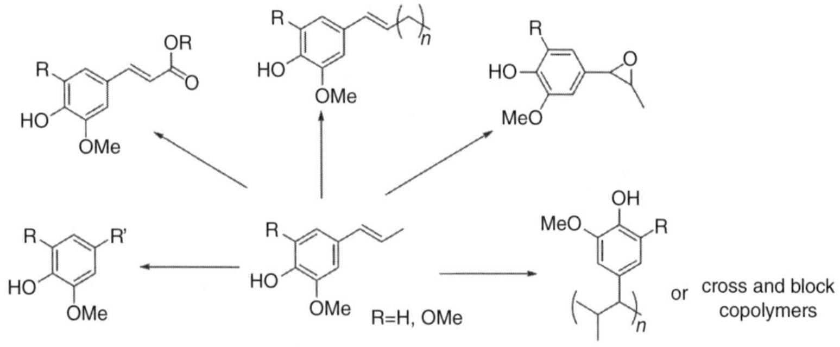

Figure 7.9 Possible applications of lignin-derived aryl propenes in the fine and bulk chemistry fields [119]. Adapted from Galkin and Samec (2014) *ChemSusChem*, **7**, 2154–2158. Reproduced with permission of Wiley-VCH Verlag.

7.6 Stability and Toxicity of Alkylphenols

In order to function as solvents, the stability of alkylphenols is critical. Alkylphenols contain three reactive functional groups, a hydroxyl group, an alkyl group and a benzene ring, which can cause side reactions and result in their destruction when used as solvents in catalytic reactions. For example, the benzene ring can undergo hydrogenation in the presence of hydrogenation catalysts under a hydrogen atmosphere. This has been demonstrated in the hydrogenation of levulinic acid to GVL [128]. But this case can be solved by modifying the hydrogenation catalysts. Under acidic conditions, alkylphenols can isomerize, dealkylate and transalkylate. Even without acid catalysts, these reactions are able to proceed in water under severe reaction conditions (like in supercritical water) [162]. Also, alkylation of the benzene ring with substrates, reagents or solvents (especially with alcohols and double bonds) or products (like HMF) can occur [163].

Lignin-derived alkylphenols have an additional methoxyl group. Apart from the above-mentioned reactions, the O–CH$_3$ bond can be hydrolysed under severe reaction conditions. Hydrogenation reactions can also occur on the methoxyl group. This has been demonstrated in the hydrogenation of levulinic acid to GVL [109]. Transalkylation of methoxyl group is observed under thermal conditions [164]. Most importantly, as described in section 7.2, alkylphenols are sensitive to oxygen.

In order to implement alkylphenols as organic solvents, it is important to consider the environmental health and safety issues. Alkylphenols are hazardous substances and dangerous, both to people and to the environment, if handled improperly. It was reported that the toxicities of these alkylphenol solvents (i.e. SBP, NHP and PG) are in the range of many organic solvents used in the chemical industry, such as aromatics, halogenated hydrocarbons and ethers [131]. The high boiling points of these alkylphenols facilitate their handling compared to some common solvents.

The environmental safety of alkylphenols strongly depends on the chemical structure of these compounds. Traditional – non-lignin-based – alkylphenols with long alkyl chains are very hydrophobic and only slightly solubilize in water. As a result, during wastewater treatment, these compounds tend to adsorb strongly in soils, sediments and sludges. The degradation is slow, with half-lives of multiple weeks to years [30, 32]. Their persistence in the environment can be disastrous for aquatic organisms, since alkylphenols are highly toxic (acute and chronic) for these species. In addition, alkylphenols are able to bio-accumulate in their tissues, wherein these compounds are found to be oestrogen-mimicking by interference with the binding of natural oestrogens. As a result, the intake of alkylphenols from fish also raises concerns for human health [165–167]. Owing to the negative effect on health and the environment, the production and use of nonylphenol and nonylphenol ethoxylates have been prohibited by the European Union [168]. The presence of an additional methoxy group on the phenolic ring significantly diminishes the binding affinity of phenolic structures for the oestrogen receptor [169, 170]. From this point of view, bio-derived methoxylated alkylphenols are promising alternatives to traditional alkylphenols as their toxicity is significantly lower [171]. Furthermore, methoxylated alkylphenols from lignin can possess an unsaturated alkyl chain, e.g. *iso*-eugenol or eugenol. The unsaturation is also proposed to benefit the biodegradability of the alkylphenol, as unsaturated compounds often degrade faster in various environments than their saturated counterparts [172, 173].

7.7 Conclusions and Perspectives

The development of bio-based solvents is an important step towards a sustainable chemical industry. Within the wide variety of bio-based solvents, alkylphenols, and in particular the lignin-derived methoxylated alkylphenols, show a considerable potential as solvents for future biorefineries. This is not only because the synthesis of these compounds is bio-based, but also because they are better degradable and less toxic than traditional alkylphenols from Friedel–Crafts alkylation. Moreover, by implementation of lignin-derived alkylphenols as solvents, the purchase and transport of large amounts of some non-bio-based solvents for biorefineries will be reduced considerably, which can significantly improve the economics of the entire biorefinery process. However, many opportunities still exist to extend the application of alkylphenols as solvents. Although a number of technologies for producing alkylphenols from biomass (especially lignin) have been developed, there is still room for improvement in the efficiency and economics of these processes. Moreover, new bio-based synthesis routes to long-chain alkylphenols (alkyl chains > C_5) are definitely worth exploring. Such compounds exhibit other properties and are used in other applications than short-chain alkylphenols, which are directly derived from lignin. In order to use alkylphenols on a large scale, the (long-term) stability, toxicity and environmental impact should be thoroughly investigated. Finally, utilization of alkylphenols as organic solvents in other fields requires further exploration.

Acknowledgements

Y.L. acknowledges funding from the China Scholarship Council (CSC) for a doctoral fellowship (201404910467). A.D and D.V. are grateful for funding from the Research Foundation Flanders (FWO) for a doctoral and a post-doctoral fellowship, respectively. FISCH-ARBOREF financial support by the Flemish government is acknowledged.

References

1. Kamm, B., Gruber, P.R. and Kamm, M. (2007) Biorefineries – industrial processes and products. In *Ullmann's Encyclopedia of Industrial Chemistry*, Wiley-VCH, Weinheim, pp. 1–38.
2. Nicholas, K.M. (ed.) (2014) *Selective Catalysis for Renewable Feedstocks and Chemicals*, Springer, Switzerland.
3. Smith, B., Greenwell, H.C. and Whiting, A. (2009) Catalytic upgrading of tri-glycerides and fatty acids to transport biofuels. *Energy Environ. Sci.*, **2**, 262–271.
4. van Putten, R.J., van der Waal, J.C., De Jong, E., et al. (2013) Hydroxymethylfurfural, a versatile platform chemical made from renewable resources. *Chem. Rev.*, **113**, 1499–1597.
5. Cai, C.M., Zhang, T., Kumar, R. and Wyman, C.E. (2014) Integrated furfural production as a renewable fuel and chemical platform from lignocellulosic biomass. *J. Chem. Technol. Biotechnol.*, **89**, 2–10.
6. Díaz-Álvarez, A.E., Francos, J., Lastra-Barreira, B., et al. (2011) Glycerol and derived solvents: new sustainable reaction media for organic synthesis. *Chem. Commun.*, **47**, 6208–6227.
7. Gu, Y. and Jérôme, F. (2013) Bio-based solvents: an emerging generation of fluids for the design of eco-efficient processes in catalysis and organic chemistry. *Chem. Soc. Rev.*, **42**, 9550–9570.
8. García, J.I., García-Marín, H. and Pires, E. (2014) Glycerol based solvents: synthesis, properties and applications. *Green Chem.*, **16**, 1007–1033.
9. Gu, Y. and Jérôme, F. (2010) Glycerol as a sustainable solvent for green chemistry. *Green Chem.*, **12**, 1127–1138.
10. Pace, V., Hoyos, P., Castoldi, L., *et al.* (2012) 2-Methyltetrahydrofuran (2-MeTHF): a biomass-derived solvent with broad application in organic chemistry. *ChemSusChem*, **5**, 1369–1379.
11. Zhang, Z. (2016) Synthesis of γ-valerolactone from carbohydrates and its applications. *ChemSusChem*, **9**, 156–171.
12. Alonso, D.M., Wettstein, S.G. and Dumesic, J.A. (2013) Gamma-valerolactone, a sustainable platform molecule derived from lignocellulosic biomass. *Green Chem.*, **15**, 584–595.
13. Yang, J., Tan, J. and Gu, Y. (2012) Lactic acid as an invaluable bio-based solvent for organic reactions. *Green Chem.*, **14**, 3304–3317.
14. Pereira, C.S., Silva, V.M. and Rodrigues, A.E. (2011) Ethyl lactate as a solvent: properties, applications and production processes – a review. *Green Chem.*, **13**, 2658–2671.
15. Mäki-Arvela, P., Simakova, I.L., Salmi, T. and Murzin, D.Y. (2013) Production of lactic acid/lactates from biomass and their catalytic transformations to commodities. *Chem. Rev.*, **114**, 1909–1971.
16. Zakzeski, J., Bruijnincx, P.C., Jongerius, A.L. and Weckhuysen, B.M. (2010) The catalytic valorization of lignin for the production of renewable chemicals. *Chem. Rev.*, **110**, 3552–3599.
17. Li, C., Zhao, X., Wang, A., *et al.* (2015) Catalytic transformation of lignin for the production of chemicals and fuels. *Chem. Rev.*, **115**, 11559–11624.
18. Lorenc, J.F., Lambeth, G. and Scheffer, W. (1992) Alkylphenols. In *Kirk-Othmer Encyclopedia of Chemical Technology*, John Wiley & Sons, Inc., New York, pp. 113–143.

19. Fiege, H., Voges, H.W., Hamamoto, T., *et al.* (2000) Phenol derivatives. In *Ullmann's Encyclopedia of Industrial Chemistry*, Wiley-VCH, Weinheim, pp. 521–582.
20. Tokyo Chemical Industry. www.tcichemicals.com.
21. Sigma-Aldrich. www.sigmaaldrich.com.
22. Alfa Aesar. www.alfa.com.
23. PubChem, National Center for Biotechnology Information. pubchem.ncbi.nlm.nih.gov.
24. ChemSpider. www.chemspider.com.
25. NIST, National Institute of Standards and Technology. www.nist.gov.
26. The Good Scents Company Information System. www.thegoodscentscompany.com.
27. Yalkowsky, S.H., He, Y. and Jain, P. (eds) (2010) *Handbook of Aqueous Solubility Data*, 2nd edn, CRC Press, Boca Raton, FL.
28. Lambropoulou, D.A. and Nollet, L.M. (eds) (2014) *Transformation Products of Emerging Contaminants in the Environment: Analysis, Processes, Occurrence, Effects and Risks*, John Wiley & Sons, Ltd, Chichester.
29. Environment Agency (2008) *Environmental Risk Evaluation Report: 4-tert-Pentylphenol* (CAS No. 80–46–6), Environment Agency, Bristol.
30. Environment Agency (2005) Environmental Risk Evaluation Report: 4-tert-Octylphenol, Environment Agency, Bristol.
31. Environment Agency (2007) Environmental Risk Evaluation Report: para-C_{12}-Alkyl phenols (Dodecylphenol and Tetrapropenylphenol), Environment Agency, Bristol.
32. European Chemicals Bureau (2002) European Union Risk Assessment Report: 4-Nonylphenol (Branched) and nonylphenol. Final Report. European Commission, Luxembourg.
33. Verboekend, D., Liao, Y., Schutyser, W. and Sels, B.F. (2016) Alkylphenols to phenol and olefins by zeolite catalysis: a pathway to valorize raw and fossilized lignocellulose. *Green Chem.*, **18**, 297–306.
34. Schmidt, R.J. (2005) Industrial catalytic processes – phenol production. *Appl. Catal. A: Gen.*, **280**, 89–103.
35. Wei, L., Shang, Y. and Yang, P. (2008) Alkylation of phenol with *iso*propanol over MCM-49 zeolites. *React. Kinet. Catal. Lett.*, **93**, 265–271.
36. Zhang, K., Xiang, S., Zhang, H., *et al.* (2002) Phenol alkylation with *tert*-butyl alcohol catalyzed by HM zeolite. *React. Kinet. Catal. Lett.*, **77**, 13–19.
37. Zhang, K., Zhang, H., Xu, G., *et al.* (2001) Alkylation of phenol with *tert*-butyl alcohol catalyzed by large pore zeolites. *Appl. Catal. A: Gen.*, **207**, 183–190.
38. Chaudhuri, B. and Sharma, M.M. (1991) Alkylation of phenol with *alpha*-methylstyrene, propylene, butenes, *iso*amylene, 1-octene, and diisobutylene: heterogeneous vs. homogeneous catalysts. *Ind. Eng. Chem. Res.*, **30**, 227–231.
39. Velu, S. and Swamy, C. (1996) Alkylation of phenol with 1-propanol and 2-propanol over catalysts derived from hydrotalcite-like anionic clays. *Catal. Lett.*, **40**, 265–272.
40. Balasubramanian, V., Umamaheshwari, V., Kumar, I.S., *et al.* (1988) Alkylation of phenol with methanol over ion-exchanged Y-zeolites, *Proc. Indian Acad. Sci. – Chem. Sci.*, **111**, 453–460.
41. Krishnan, V., Ojha, K. and Pradhan, N.C. (2002) Alkylation of phenol with tertiary butyl alcohol over zeolites. *Org. Process Res. Dev.*, **6**, 132–137.
42. Sad, M.E., Duarte, H.A., Padró, C.L. and Apesteguía, C.R. (2014) Selective synthesis of *p*-ethylphenol by gas-phase alkylation of phenol with ethanol. *Appl. Catal. A: Gen.*, **486**, 77–84.

43. Das, J. and Halgeri, A.B. (2000) Selective synthesis of *para*-ethylphenol over pore size tailored zeolite. *Appl. Catal. A: Gen.*, **194**, 359–363.

44. Balsama, S., Beltrame, P., Beltrame, P.L., *et al.* (1984) Alkylation of phenol with methanol over zeolites. *Appl. Catal.*, **13**, 161–170.

45. Xu, J., Yan, A.Z. and Xu, Q.H. (1997) Alkylation of phenol with methanol on H-beta zeolite. *React. Kinet. Catal. Lett.*, **62**, 71–74.

46. Xu, W., Miller, S.J., Agrawal, P.K. and Jones, C.W. (2013) Zeolite topology effects in the alkylation of phenol with propylene. *Appl. Catal. A: Gen.*, **459**, 114–120.

47. Wang, B., Lee, C.W., Cai, T.X. and Park, S.E. (2001) Identification and influence of acidity on alkylation of phenol with propylene over ZSM-5. *Catal. Lett.*, **76**, 219–224.

48. Chang, N.S., Chen, C.C., Chu, S.J., et al. (1989) Acidity effect of ZSM-5 zeolites on phenol methylation reaction. *Stud. Surf. Sci. Catal.*, **46**, 223–230.

49. Bregolato, M., Bolis, V., Busco, C., et al. (2007) Methylation of phenol over high-silica beta zeolite: effect of zeolite acidity and crystal size on catalyst behaviour. *J. Catal.*, **245**, 285–300.

50. Dumitriu, E. and Hulea, V. (2003) Effects of channel structures and acid properties of large-pore zeolites in the liquid-phase *tert*-butylation of phenol. *J. Catal.*, **218**, 249–257.

51. Sad, M.E., Padró, C.L. and Apesteguía, C.R. (2010) Study of the phenol methylation mechanism on zeolites HBEA, HZSM5 and HMCM22. *J. Mol. Catal. A: Chem.*, **327**, 63–72.

52. Bal, R., Tope, B.B. and Sivasanker, S. (2002) Vapour phase *O*-methylation of dihydroxy benzenes with methanol over cesium-loaded silica, a solid base. *J. Mol. Catal. A: Chem.*, **181**, 161–171.

53. Velu, S. and Swamy, C.S. (1996) Selective *C*-alkylation of phenol with methanol over catalysts derived from copper–aluminium hydrotalcite-like compounds. *Appl. Catal. A: Gen.*, **145**, 141–153.

54. Rao, V.V., Kumari, V.D. and Narayanan, S. (1989) Selective alkylation of phenol to 2,6-xylenol over vanadia–chromia mixed oxide catalysts. *Appl. Catal.*, **49**, 165–174.

55. Velu, S. and Swamy, C.S. (1997) Effect of substitution of Fe^{3+}/Cr^{3+} on the alkylation of phenol with methanol over magnesium–aluminium calcined hydrotalcite. *Appl. Catal. A: Gen.*, **162**, 81–91.

56. Bolognini, M., Cavani, F., Scagliarini, D., *et al.* (2002) Heterogeneous basic catalysts as alternatives to homogeneous catalysts: reactivity of Mg/Al mixed oxides in the alkylation of *m*-cresol with methanol. *Catal. Today*, **75**, 103–111.

57. Sato, S., Takahashi, R., Sodesawa, T., *et al.* (1999) *Ortho*-selective alkylation of phenol with 1-propanol catalyzed by CeO_2–MgO. *J. Catal.*, **184**, 180–188.

58. Shi, Q., Pan, N., Long, H., *et al.* (2012) Characterization of middle-temperature gasification coal tar. Part 3: molecular composition of acidic compounds. *Energy Fuels*, **27**, 108–117.

59. Long, H., Shi, Q., Pan, N., et al. (2012) Characterization of middle-temperature gasification coal tar. Part 2: neutral fraction by extrography followed by gas chromatography–mass spectrometry and electrospray ionization coupled with Fourier transform ion cyclotron resonance mass spectrometry. *Energy Fuels*, **26**, 3424–3431.

60. Pan, N., Cui, D., Li, R., *et al.* (2012) Characterization of middle-temperature gasification coal tar. Part 1: bulk properties and molecular compositions of distillates and basic fractions. *Energy Fuels*, **26**, 5719–5728.

61. Mochida, I., Okuma, O. and Yoon, S. H. (2013) Chemicals from direct coal liquefaction. *Chem. Rev.*, **114**, 1637–1672.

62. Granda, M., Blanco, C., Alvarez, P., *et al.* (2013) Chemicals from coal coking. *Chem. Rev.*, **114**, 1608–1636.

63. Omais, B., Courtiade, M., Charon, N., *et al.* (2010) Characterization of oxygenated species in coal liquefaction products: an overview. *Energy Fuels*, **24**, 5807–5816.

64. Gogotov, A. F., Levchuk, A.A., Tai D.C., et al. (2015) Coal tar phenols as a promising intermediate for producing high-performance polymerization inhibitors in petrochemical industry. *Petrol. Chem.*, **55**, 530–536.

65. Weber, M., Weber, M. and Kleine-Boymann, M. (2004) Phenol. In *Ullmann's Encyclopedia of Industrial Chemistry*, Wiley-VCH, Weinheim, pp. 503–519.

66. Michałowicz, J. and Duda, W. (2007) Phenols – sources and toxicity. *Pol. J. Environ. Stud.*, **16**, 347–362.

67. Fiege, H. (2000) Cresols and xylenols. In *Ullmann's Encyclopedia of Industrial Chemistry*, Wiley-VCH, Weinheim, pp. 419–461.

68. Cislak, F.E. and Otto, M.M. (1946) Process for separating *para*-ethyl phenol from *meta*-ethyl phenol. US Patent 2490670, 19 July 1946.

69. Pandey, M.P. and Kim, C.S. (2011) Lignin depolymerization and conversion: a review of thermochemical methods. *Chem. Eng. Technol.*, **34**, 29–41.

70. Xu, C., Arancon, R.A.D., Labidi, J. and Luque, R. (2014) Lignin depolymerisation strategies: towards valuable chemicals and fuels. *Chem. Soc. Rev.*, **43**, 7485–7500.

71. Azadi, P., Inderwildi, O.R., Farnood, R. and King, D.A. (2013) Liquid fuels, hydrogen and chemicals from lignin: a critical review. *Renew. Sustain. Energy Rev.*, **21**, 506–523.

72. Ma, R., Xu, Y. and Zhang, X. (2015) Catalytic oxidation of biorefinery lignin to value-added chemicals to support sustainable biofuel production. *ChemSusChem*, **8**, 24–51.

73. Das, L., Kolar, P. and Sharma-Shivappa, R. (2012) Heterogeneous catalytic oxidation of lignin into value-added chemicals. *Biofuels*, **3**, 155–166.

74. Behling, R., Valange, S. and Chatel, G. (2016) Heterogeneous catalytic oxidation for lignin valorization into valuable chemicals: What results? What limitations? What trends? *Green Chem.*, **18**, 1839–1854.

75. Mu, W., Ben, H., Ragauskas, A. and Deng, Y. (2013) Lignin pyrolysis components and upgrading – technology review. *BioEnergy Res.*, **6**, 1183–1204.

76. Johnson, D.K., Chornet, E., Zmierczak, W. and Shabtai, J. (2002) Conversion of lignin into a hydrocarbon product for blending with gasoline. *ACS Div. Fuel Chem. Preprints*, **47**, 380–381.

77. Shabtai, J.S., Zmierczak, W.W. and Chornet, E. (1998) Process for conversion of lignin to reformulated hydrocarbon gasoline. US Patent 5959167, 19 August 1998.

78. Shabtai, J.S., Zmierczak, W.W. and Chornet, E. (1999) Process for conversion of lignin to reformulated, partially oxygenated gasoline. US Patent 6172272 B1, 18 August 1999.

79. Shabtai, J.S., Zmierczak, W.W., Chornet, E. and Johnson, D. (2001) Process for converting lignins into a high octane additive. US Patent 2003/0100807 A1, 5 October 2001.

80. Vigneault, A., Johnson, D.K. and Chornet, E. (2007) Base-catalyzed depolymerization of lignin: separation of monomers. *Can. J. Chem. Eng.*, **85**, 906–916.

81. Miller, J.E., Evans, L., Littlewolf, A. and Trudell, D.E. (1999) Batch microreactor studies of lignin and lignin model compound depolymerization by bases in alcohol solvents. *Fuel*, **78**, 1363–1366.

82. Roberts, V.M., Stein, V., Reiner, T., *et al.* (2011) Towards quantitative catalytic lignin depolymerization. *Chemistry, Eur. J.*, **17**, 5939–5948.

83. Toledano, A., Serrano, L. and Labidi, J. (2012) Organosolv lignin depolymerization with different base catalysts. *J. Chem. Technol. Biotechnol.*, **87**, 1593–1599.

84. Toledano, A., Serrano, L. and Labidi, J. (2014) Improving base catalyzed lignin depolymerization by avoiding lignin repolymerization. *Fuel*, **116**, 617–624.

85. Lavoie, J.M., Baré, W. and Bilodeau, M. (2011) Depolymerization of steam-treated lignin for the production of green chemicals. *Bioresour. Technol.*, **102**, 4917–4920.

86. Katahira, R., Mittal, A., McKinney, K., *et al.* (2016) Base-catalyzed depolymerization of biorefinery lignins. *ACS Sustain. Chem. Eng.*, **4**, 1474–1486.

87. Sturgeon, M.R., O'Brien, M.H., Ciesielski, P.N., *et al.* (2014) Lignin depolymerisation by nickel supported layered-double hydroxide catalysts. *Green Chem.*, **16**, 824–835.

88. Kruger, J.S., Cleveland, N.S., Zhang, S., *et al.* (2016) Lignin depolymerization with nitrate-intercalated hydrotalcite catalysts. *ACS Catal.*, **6**, 1316–1328.

89. Barta, K., Matson, T.D., Fettig, M.L., *et al.* (2010) Catalytic disassembly of an Organosolv lignin via hydrogen transfer from supercritical methanol. *Green Chem.*, **12**, 1640–1647.

90. Barta, K., Warner, G.R., Beach, E.S. and Anastas, P.T. (2014) Depolymerization of Organosolv lignin to aromatic compounds over Cu-doped porous metal oxides. *Green Chem.*, **16**, 191–196.

91. Matson, T.D., Barta, K., Iretskii, A. V. and Ford, P.C. (2011) One-pot catalytic conversion of cellulose and of woody biomass solids to liquid fuels. *J. Am. Chem. Soc.*, **133**, 14090–14097.

92. Huang, X., Korányi, T.I., Boot, M.D. and Hensen, E.J.M. (2014) Catalytic depolymerization of lignin in supercritical ethanol. *ChemSusChem*, **7**, 2276–2288.

93. Huang, X., Korányi, T.I., Boot, M.D. and Hensen, E.J.M. (2015) Ethanol as capping agent and formaldehyde scavenger for efficient depolymerization of lignin to aromatics. *Green Chem.*, **17**, 4941–4950.

94. Ma, R., Hao, W., Ma, X., *et al.* (2014) Catalytic ethanolysis of Kraft lignin into high-value small-molecular chemicals over a nanostructured α-molybdenum carbide catalyst. *Angew. Chem., Int. Edn*, **53**, 7310–7315.

95. Zakzeski, J., Jongerius, A.L., Bruijnincx, P.C. and Weckhuysen, B.M. (2012) Catalytic lignin valorization process for the production of aromatic chemicals and hydrogen. *ChemSusChem*, **5**, 1602–1609.

96. Xu, W., Miller, S.J., Agrawal, P.K. and Jones, C.W. (2012) Depolymerization and hydrodeoxygenation of switchgrass lignin with formic acid. *ChemSusChem*, **5**, 667–675.

97. Onwudili, J.A. and Williams, P.T. (2014) Catalytic depolymerization of alkali lignin in subcritical water: influence of formic acid and Pd/C catalyst on the yields of liquid monomeric aromatic products. *Green Chem.*, **16**, 4740–4748.

98. Song, Q., Wang, F. and Xu, J. (2012) Hydrogenolysis of lignosulfonate into phenols over heterogeneous nickel catalysts. *Chem. Commun.*, **48**, 7019–7021.

99. Kasakov, S., Shi, H., Camaioni, D.M., *et al.* (2015) Reductive deconstruction of Organosolv lignin catalyzed by zeolite supported nickel nanoparticles. *Green Chem.*, **17**, 5079–5090.

100. Molinari, V., Clavel, G., Graglia, M., *et al.* (2016) Mild continuous hydrogenolysis of Kraft lignin over titanium nitride–nickel catalyst. *ACS Catal.*, **6**, 1663–1670.

101. Bouxin, F.P., McVeigh, A., Tran, F., *et al.* (2015) Catalytic depolymerisation of isolated lignins to fine chemicals using a Pt/alumina catalyst: Part 1 – impact of the lignin structure. *Green Chem.*, **17**, 1235–1242.

102. Zhang, J., Teo, J., Chen, X., *et al.* (2014) A series of NiM (M = Ru, Rh, and Pd) bimetallic catalysts for effective lignin hydrogenolysis in water. *ACS Catal.*, **4**, 1574–1583.

103. Zhang, J., Asakura, H., van Rijn, J., *et al.* (2014) Highly efficient, NiAu-catalyzed hydrogenolysis of lignin into phenolic chemicals. *Green Chem.*, **16**, 2432–2437.

104. Narani, A., Chowdari, R.K., Cannilla, C., *et al.* (2015) Efficient catalytic hydrotreatment of Kraft lignin to alkylphenolics using supported NiW and NiMo catalysts in supercritical methanol. *Green Chem.*, **17**, 5046–5057.

105. Kloekhorst, A. and Heeres, H.J. (2015) Catalytic hydrotreatment of Alcell lignin using supported Ru, Pd, and Cu catalysts. *ACS Sustain. Chem. Eng.*, **3**, 1905–1914.

106. Kloekhorst, A., Wildschut, J. and Heeres, H.J. (2014) Catalytic hydrotreatment of pyrolytic lignins to give alkylphenolics and aromatics using a supported Ru catalyst. *Catal. Sci. Technol.*, **4**, 2367–2377.

107. Kumar, C.R., Anand, N., Kloekhorst, A., *et al.* (2015) Solvent free depolymerization of Kraft lignin to alkyl-phenolics using supported NiMo and CoMo catalysts. *Green Chem.*, **17**, 4921–4930.

108. Yan, N., Zhao, C., Dyson, P.J., *et al.* (2008) Selective degradation of wood lignin over noble-metal catalysts in a two-step process. *ChemSusChem*, **1**, 626–629.

109. Azadi, P., Carrasquillo-Flores, R., Pagán-Torres, Y.J., *et al.* (2012) Catalytic conversion of biomass using solvents derived from lignin. *Green Chem.*, **14**, 1573–1576.

110. Li, C., Zheng, M., Wang, A. and Zhang, T. (2012) One-pot catalytic hydrocracking of raw woody biomass into chemicals over supported carbide catalysts: simultaneous conversion of cellulose, hemicellulose and lignin. *Energy Environ. Sci.*, **5** (4), 6383–6390.

111. Song, Q., Wang, F., Cai, J., *et al.* (2013) Lignin depolymerization (LDP) in alcohol over nickel-based catalysts via a fragmentation–hydrogenolysis process. *Energy Environ. Sci.*, **6**, 994–1007.

112. Parsell, T., Yohe, S., Degenstein, J., *et al.* (2015) A synergistic biorefinery based on catalytic conversion of lignin prior to cellulose starting from lignocellulosic biomass. *Green Chem.*, **17**, 1492–1499.

113. Luo, H., Klein, I.M., Jiang, Y., *et al.* (2016) Total utilization of *Miscanthus* biomass, lignin and carbohydrates, using earth abundant nickel catalyst. *ACS Sustain. Chem. Eng.*, **4**, 2316–2322.

114. Van den Bosch, S., Schutyser, W., Vanholme, R., *et al.* (2015) Reductive lignocellulose fractionation into soluble lignin-derived phenolic monomers and dimers and processable carbohydrate pulps. *Energy Environ. Sci.*, **8**, 1748–1763.

115. Van den Bosch, S., Schutyser, W., Koelewijn, S.F., *et al.* (2015) Tuning the lignin oil OH-content with Ru and Pd catalysts during lignin hydrogenolysis on birch wood. *Chem. Commun.*, **51**, 13158–13161.

116. Renders, T., Schutyser, W., Van den Bosch, S., *et al.* (2016) Influence of acidic (H_3PO_4) and alkaline (NaOH) additives on the catalytic reductive fractionation of lignocellulose. *ACS Catal.*, **6**, 2055–2066.

117. Schutyser, W., Van den Bosch, S., Renders, T., *et al.* (2015) Influence of bio-based solvents on the catalytic reductive fractionation of birch wood. *Green Chem.*, **17**, 5035–5045.

118. Ferrini, P. and Rinaldi, R. (2014) Catalytic biorefining of plant biomass to non-pyrolytic lignin bio-oil and carbohydrates through hydrogen transfer reactions. *Angew. Chem., Int. Edn*, **53**, 8634–8639.

119. Galkin, M.V. and Samec, J.S. (2014) Selective route to 2-propenyl aryls directly from wood by a tandem Organosolv and palladium-catalysed transfer hydrogenolysis. *ChemSusChem*, **7**, 2154–2158.

120. Jiang, Z., He, T., Li, J. and Hu, C. (2014) Selective conversion of lignin in corncob residue to monophenols with high yield and selectivity. *Green Chem.*, **16**, 4257–4265.

121. Hu, L., Luo, Y., Cai, B., *et al.* (2014) The degradation of the lignin in *Phyllostachys heterocycla* cv. *pubescens* in an ethanol solvothermal system. *Green Chem.*, **16**, 3107–3116.

122. Saidi, M., Samimi, F., Karimipourfard, D., *et al.* (2014) Upgrading of lignin-derived bio-oils by catalytic hydrodeoxygenation. *Energy Environ. Sci.*, **7**, 103–129.

123. Vuori, A. and Bredenberg, J.B. (1984) Hydrogenolysis and hydrocracking of the carbon–oxygen bond. 5. Hydrogenolysis of 4-propylguaiacol by sulfided CoO-MoO_3/γ-Al_2O_3. *Holzforschung – Int. J. Biol., Chem., Phys. Technol. Wood*, **38**, 253–262.

124. Ratcliff, M.A., Johnson, D.K., Posey, F.L., *et al.* (1988) Hydrodeoxygenation of a lignin model compound. In *Research in Thermochemical Biomass Conversion* (eds A.V. Bridgwater and J.L. Kuester), Elsevier Applied Science, London, pp. 941–955.

125. Joshi, N. and Lawal, A. (2013) Hydrodeoxygenation of 4-propylguaiacol (2-methoxy-4-propylphenol) in a microreactor: performance and kinetic studies. *Ind. Eng. Chem. Res.*, **52**, 4049–4058.

126. Alonso, D.M., Wettstein, S.G., Bond, J.Q., *et al.* (2011) Production of biofuels from cellulose and corn stover using alkylphenol solvents. *ChemSusChem*, **4**, 1078–1081.

127. Dumesic, J., Bond, J., Alonso, D. and Root, T. (2011) Method to produce and recover levulinic acid and/or gamma-valerolactone from aqueous solutions using alkylphenols. US Patent 8389761 B2, 25 May 2011.

128. Wettstein, S.G., Bond, J.Q., Alonso, D.M., *et al.* (2012) RuSn bimetallic catalysts for selective hydrogenation of levulinic acid to γ-valerolactone. *Appl. Catal. B: Environ.*, **117**, 321–329.

129. Gürbüz, E.I., Wettstein, S.G. and Dumesic, J.A. (2012) Conversion of hemicellulose to furfural and levulinic acid using biphasic reactors with alkylphenol solvents. *ChemSusChem*, **5**, 383–387.

130. Dumesic, J., Alonso, D., Bond, J., *et al.* (2011) Method to produce, recover and convert furan derivatives from aqueous solutions using alkylphenol extraction. US Patent 8389749 B2, 25 May 2011.

131. Gürbüz, E.I. (2012) Strategies for the catalytic conversion of lignocellulose-derived carbohydrates to chemicals and fuels. Ph.D. Thesis. University of Wisconsin–Madison.

132. Lin, L., Ma, S., Li, P., *et al.* (2015) Mutual solubilities for the water-2-*sec*-butylphenol system and partition coefficients for furfural and formic acid in the water-2-*sec*-butylphenol system. *J. Chem. Eng. Data*, **60**, 1926–1933.

133. Pagan-Torres, Y.J., Wang, T., Gallo, J.M.R., *et al.* (2012) Production of 5-hydroxymethyl furfural from glucose using a combination of Lewis and Brønsted acid catalysts in water in a biphasic reactor with an alkylphenol solvent. *ACS Catal.*, **2**, 930–934.

134. Dumesic, J.A., Pagán-Torres, Y.J., Wang, T. and Shanks, B.H. (2011) Lewis and Bronsted–Lowry acid-catalyzed production 5-hydroxymethylfurfural (HMF) from glucose. US Patent 8642791 B1, 7 December 2011.

135. Carrasquillo-Flores, R., Käldström, M., Schüth, F., *et al.* (2013) Mechanocatalytic depolymerization of dry (ligno)cellulose as an entry process for high-yield production of furfurals. *ACS Catal.*, **3**, 993–997.

136. Luterbacher, J.S., Rand, J.M., Alonso, D.M., *et al.* (2014) Nonenzymatic sugar production from biomass using biomass-derived γ-valerolactone. *Science*, **343**, 277–280.

137. Luterbacher, J.S., Alonso, D.M., Rand, J.M., *et al.* (2015) Solvent-enabled nonenyzmatic sugar production from biomass for chemical and biological upgrading. *ChemSusChem*, **8**, 1317–1322.

138. Dumesic, J.A. and Luterbacher, J.S. (2013) Method to produce water-soluble sugars from biomass using solvents containing lactones. US Patent 9045804 B2, 8 January 2013.

139. Blumenthal, L.C., Jens, C.M., Ulbrich, J., *et al.* (2016) Systematic identification of solvents optimal for the extraction of 5-hydroxymethylfurfural from aqueous reactive solutions. *ACS Sustain. Chem. Eng.*, **4**, 228–235.

140. Kim, S. and Han, J. (2016) A catalytic biofuel production strategy involving separate conversion of hemicellulose and cellulose using 2-*sec*-butylphenol (SBP) and lignin-derived (LD) alkylphenol solvents. *Bioresour. Technol.*, **204**, 1–8.

141. Byun, J. and Han, J. (2016) Catalytic production of biofuels (butene oligomers) and biochemicals (tetrahydrofurfuryl alcohol) from corn stover. *Bioresour. Technol.*, **211**, 360–366.

142. Byun, J. and Han, J. (2016) Process synthesis and analysis for catalytic conversion of lignocellulosic biomass to fuels: separate conversion of cellulose and hemicellulose using 2-*sec*-butylphenol (SBP) solvent. *Appl. Energy*, **171**, 483–490.

143. Han, J., Sen, S.M., Luterbacher, J.S., *et al.* (2015) Process systems engineering studies for the synthesis of catalytic biomass-to-fuels strategies. *Comput. Chem. Eng.*, **81**, 57–69.

144. Okuda, K., Man, X., Umetsu, M., *et al.* (2004) Efficient conversion of lignin into single chemical species by solvothermal reaction in water–*p*-cresol solvent. *J. Phys.: Condens. Matter*, **16**, S1325-S1330.

145. Okuda, K., Ohara, S., Umetsu, M., *et al.* (2008) Disassembly of lignin and chemical recovery in supercritical water and *p*-cresol mixture: studies on lignin model compounds. *Bioresour. Technol.*, **99**, 1846–1852.

146. Birajdar, S.D., Padmanabhan, S. and Rajagopalan, S. (2014) Rapid solvent screening using thermodynamic models for recovery of 2,3-butanediol from fermentation by liquid–liquid extraction. *J. Chem. Eng. Data*, **59**, 2456–2463.

147. Khuspe, G.D., Sakhare, R.D., Navale, S.T., *et al.* (2013) Nanostructured SnO_2 thin films for NO_2 gas sensing applications. *Ceram. Int.*, **39**, 8673–8679.

148. Yuan, S., Guo, X., Aili, D., *et al.* (2014) Poly(imide benzimidazole)s for high temperature polymer electrolyte membrane fuel cells. *J. Membr. Sci.*, **454**, 351–358.

149. Chen, J.C., Wu, J.A., Chang, H.W. and Lee, C.Y. (2014) Organosoluble polyimides derived from asymmetric 2-substituted-and 2,2′,6-trisubstituted-4,4′-oxydianilines. *Polym. Int.*, **63**, 352–362.

150. Landis, A.L. and Naselow, A.B. (1985) Method of preparing high molecular weight polyimide, product and use. US Patent 4645824, 25 November 1985.

151. Ikeda, K., Endo, M. and Abe, S. (1997) *p*-Nonylphenol as a versatile solvent for the liquid–liquid extraction of cationic metal complexes. *Anal. Commun.*, **34**, 183–184.

152. Ikeda, K. and Abe, S. (1998) Liquid–liquid extraction of cationic metal complexes with *p*-nonylphenol solvent: application to crown and thiacrown ether complexes of lead(ii) and copper(ii). *Anal. Chim. Acta*, **363**, 165–170.

153. Schutyser, W., Van den Bosch, S., Dijkmans, J., *et al.* (2015) Selective nickel-catalyzed conversion of model and lignin-derived phenolic compounds to cyclohexanone-based polymer building blocks. *ChemSusChem*, **8**, 1805–1818.

154. Zhao, C., Kou, Y., Lemonidou, A.A., *et al.* (2009) Highly selective catalytic conversion of phenolic bio-oil to alkanes. *Angew. Chem.*, **121**, 4047–4050.

155. Chen, M.Y., Huang, Y.B., Pang, H., *et al.* (2015) Hydrodeoxygenation of lignin-derived phenols into alkanes over carbon nanotube supported Ru catalysts in biphasic systems. *Green Chem.*, **17**, 1710–1717.

156. Huang, Y.B., Yan, L., Chen, M.Y., *et al.* (2015) Selective hydrogenolysis of phenols and phenyl ethers to arenes through direct C–O cleavage over ruthenium–tungsten bifunctional catalysts. *Green Chem.*, **17**, 3010–3017.

157. Goodyear Chemicals, Goodyear Tire and Rubber Company (1982) *Index of Commercial Antioxidants and Antiozonants*. Goodyear Chemicals.

158. Takahashi, H., Morozumi, N., Takanezawa, S. and Nakao, K. (1986) Adhesive clad insulating substrate used for producing printed circuit boards. US Patent 4837086, 31 January 1986.

159. Brooke, D., Crookes M., Johnson I., *et al.* (2005) *Prioritisation of Alkylphenols for Environmental Risk Assessment*, Environment Agency, Bristol.

160. SI Group. Global markets segments. www.siigroup.com/marketsegments.asp.

161. Chafetz, H. and Haugen, H. (1971) Sulfurized calcium alkylphenolate lubricants. US Patent 3761414, 15 September 1971.

162. Sato, T., Adschiri, T. and Arai, K. (2003) Decomposition kinetics of 2-propylphenol in supercritical water. *J. Anal. Appl. Pyrol.*, **70**, 735–746.

163. Arias, K.S., Climent, M.J., Corma, A. and Iborra, S. (2015) Synthesis of high quality alkyl naphthenic kerosene by reacting an oil refinery with a biomass refinery stream. *Energy Environ. Sci.*, **8**, 317–331.

164. Masuku, C.P., Vuori, A. and Bredenberg J.B. (1988) Thermal reactions of the bonds in lignin. I. Thermolysis of 4-propylguaiacol. *Holzforschung – Int. J. Biol., Chem., Phys. Technol. Wood*, **42**, 361–368.

165. Ekelund, R., Bergman, Å., Granmo, Å. and Berggren, M. (1990) Bioaccumulation of 4-nonylphenol in marine animals – a re-evaluation. *Environ. Pollut.*, **64**, 107–120.

166. Chen, W.L., Gwo, J.C., Wang, G.S. and Chen, C.Y. (2014) Distribution of feminizing compounds in the aquatic environment and bioaccumulation in wild tilapia tissues. *Environ. Sci. Pollut. Res.*, **21**, 11349–11360.

167. Lewis, S.K. and Lech, J.J. (1996) Uptake, disposition, and persistence of nonylphenol from water in rainbow trout (*Oncorhynchus mykiss*). *Xenobiotica*, **26**, 813–819.

168. European Commission (2003) Directive 2003/53/EC of the European Parliament and of the Council of 18 June 2003 amending for the 26th time Council Directive 76/769/EEC relating to restrictions on the marketing and use of certain dangerous substances and preparations (nonylphenol, nonylphenol ethoxylate and cement). *Official J. Eur. Union*, **178**, 24–28.

169. Martucci, C.P. and Fishman, J. (1979) Impact of continuously administered catechol estrogens on uterine growth and luteinizing hormone secretion. *Endocrinology*, **105**, 1288–1292.

170. Wang, P., McInnes, C. and Zhu, B.T. (2013) Structural characterization of the binding interactions of various endogenous estrogen metabolites with human estrogen receptor α and β subtypes: a molecular modeling study. *PLoS One*, **8**, e74615.

171. Howes, M.J.R., Houghton, P.J., Barlow, D.J., *et al.* (2002) Assessment of estrogenic activity in some common essential oil constituents. *J. Pharm. Pharmacol.*, **54**, 1521–1528.

172. Herman, D. and Roberts, D. (2006) The influence of structural components of alkyl esters on their anaerobic biodegradation in marine sediment. *Biodegradation*, **17**, 457–463.

173. American Chemistry Council's Higher Olefins Advocacy Task Group (2006) *A Comparison of the Environmental Performance of Olefin and Paraffin Synthetic Base Fluids (SBF)*, American Chemistry Council, Washington, DC.

Index

Bio-Based Solvents, First Edition. Edited by François Jérôme and Rafael Luque.
© 2017 John Wiley & Sons Ltd. Published 2017 by John Wiley & Sons Ltd.